# THE FUTURE EARTH

# THE FUTURE EARTH

## A RADICAL VISION FOR WHAT'S POSSIBLE IN THE AGE OF WARMING

ERIC HOLTHAUS

HarperCollins books may be purchased for educational, business, or sales promotional use. For information, please email the Special Markets Department at SPsales@harpercollins.com.

FIRST EDITION

*Designed by Terry McGrath*

Library of Congress Cataloging-in-Publication Data has been applied for.

ISBN 978-0-06-288316-2

20 21 22 23 24 LSC 10 9 8 7 6 5 4 3 2 1

For Roscoe and Zeke

*If I can't save us*

*then let me feel you*
*happy and safe*
*under my chin.*

*If this will drown*
*or burn*

*then let us drink starlight*
*nap under trees*
*sing on beaches—*

*the morning rush to sit indoors is for*
*what, again?*

*If we are dying*

*then let me rip open*
*and bleed Love,*
*spill it, spend it*
*see how much*
*there is*

*the reward for misers is*
*what, again?*

*If this life is ending*

*then let me begin*
*a new one*

—Lynna Odel (2019), used with permission

# CONTENTS

# PART I

# A LIVING EMERGENCY

In September 2017, Puerto Rico was just beginning to recover from one of the worst multiyear droughts in its history. Faced with severe water shortages during the height of the drought, the island's government rationed water use for two hundred thousand residents in the San Juan area, a drastic step worsened by years of austerity and colonial neglect. People there were allowed to run their taps only once every three days, forgoing one of the basic requirements of life.

Then Hurricane Maria struck, beginning the worst humanitarian crisis in modern American history.

In the span of a few hours, Hurricane Maria's 155-mile-per-hour winds and torrential rains triggered a months-long power outage that reshaped the basics of civilization in Puerto Rico. Following the storm, survivors struggled for weeks to find potable water, edible food, reliable shelter, and adequate health care. With no other options, some residents of Puerto Rico were forced to collect drinking water from toxic waste sites. Hundreds of people died because hospitals, even if they were physically accessible, didn't have electricity to provide basic services.

For survivors, Maria looked and felt like an utter reimagining of reality. Firsthand stories carried waves of shock and anguish.

In the early days after Maria, Ly Pérez, a student at the University of Puerto Rico, told me via text message that the only way she and her fellow students knew what was happening around them was by listening to the radio. "[Today's] the first time I saw pictures, and it's absolutely horrifying. They kept mentioning the word 'disaster,' and your mind would create scenarios. But in no way does it compare to the absolutely heartbreaking reality."

We have reached a point at which all weather, in every season, and in every country on Earth, is directly connected to the changes we've inflicted on our planet's atmosphere. Hurricane Maria was no exception. A 2019 study in the journal *Geophysical Research Letters* found that global warming made Maria's disastrous floods nearly five times more likely than it would have been in 1956, when high-quality rainfall record keeping began in Puerto Rico. Lead author David Keellings told the American Geophysical Union that "Maria is more extreme in its precipitation than anything else that the island has ever seen."

Hurricane Maria damaged or destroyed about 30 million trees, inflicting profound and unprecedented changes on the landscape. With the climate warming so quickly, biologists in Puerto Rico think the forests Maria destroyed will never return to their previous diversity. Many of the island's largest and slowest-growing hardwood trees, like tabonuco and balata, suffered the worst damage. Their vast canopies provide habitats for birds, bats, and tree frogs. If future hurricanes are as strong (or even stronger) than Maria, Puerto Rico's forests will eventually feature only smaller and shorter trees that are more resilient to high winds and scouring floods, which will leave local species without shelter. More than a year after the storm made landfall, satellite images showed that the island appeared definitively less green.

The storm of Hurricane Maria still hasn't let up. A full-fledged mental health crisis is ongoing throughout the island, "the largest psychosocial disaster in the United States" according to Joseph Prewitt Diaz, disaster mental health advisor for the American Red Cross. The slow recovery has created a "living emergency," a new normal that permeates daily life—characterized by despair, anxiety, and post-traumatic stress—which is more typical of refugee camps and conflict zones.

None of this was inevitable. None of this was a surprise. What is happening in Puerto Rico is the product of centuries of decisions made within a destructive system. We've known this for centuries, thanks in large part to the people whose voices were too often deemed dangerous or unworthy of our attention. Scientists are now certain that our use of fossil fuels and our destruction of the planet's ecosystems are quickly bringing the future of human civilization into doubt. My goal with this book is to help you imagine your own part in building a better world that works for everyone, regardless of status or class or gender. And to remind you that you were born at exactly the right time to help change everything.

Because we refused to take action for decades, climate change is no longer just about science. It is now, at heart, an issue of justice. The fact that *each year* we are continuing to set new record highs in greenhouse gas emissions even as our planet is rapidly warming is a shocking symptom of a larger problem in the way our society is structured. An issue of justice, climate change is also a living emergency that touches everyone and every part of society, which makes it impossible to disentangle in any meaningful way the effects of increasingly extreme weather and the unfair system that caused it. The evidence is all around us: we need to embark rapidly on a different path.

But how?

* * *

The Latin root of the word "disaster" means "ill-starred," literally a malevolent omen from the heavens. But climate-change-related disasters are no longer a matter of bad luck. We have tilted the odds toward catastrophe, particularly in the places that did the least to cause the problem. Meteorology has advanced to the point that we can now predict when and where disasters will take place. We also know that, due to the way our society is structured, the most economically and socially vulnerable parts of the planet will bear the brunt of these disasters—the people, like the survivors of Hurricane Maria in Puerto Rico, who too often experience the worst of the injustices of history.

Today, climate change compounds natural disasters, giving people less time to recover before they are plunged back into crisis mode. Residents of small islands like Puerto Rico already face limited resources for fresh water. A 2018 study found that drought in the Caribbean is increasing in severity—even as hurricanes grow stronger and downpours get more intense. And this compounding of social and climate emergencies is happening all over the world, every single year.

In 2016, the year before Maria, on the other side of the world, Cyclone Winston rapidly strengthened to the most powerful storm ever measured in the Southern Hemisphere, just hours before its landfall in Fiji. In an address to the nation following the storm, Fijian president Jioji Konrote vowed that the country would do "whatever is in its power to persuade the global community about the root cause": climate change. "This is a fight we must win," he said. "Our entire way of life is at stake." Years after landfall, as recovery drags on rainy season after rainy season, schools and families are still housed in government-issued tents.

In 2017, just a few days before Hurricane Maria hit Puerto Rico, another hurricane tore through the Caribbean. Hurricane Irma, the strongest hurricane ever to make landfall anywhere in the Atlantic Ocean, hit the island of Barbuda with winds of up to 185 miles per hour. It demolished 90 percent of the island. The entire population fled, leaving the island completely uninhabited for the first time in hundreds of years. By law and tradition, land on the island is owned communally by its residents, but in the wake of the storm, private developers are now trying to pressure the government to change the law in order to encourage more tourism.

In 2018, Typhoon Yutu hit Saipan, the largest island in the Commonwealth of the Northern Mariana Islands, a US territory.

With 180-mile-per-hour winds, it was the strongest storm in the history of the Mariana Islands. Before the storm, Saipan had been one of the world's fastest-growing tourist destinations, but its casino, a primary draw, has struggled to remain profitable since Yutu struck, and the government has had to scale back its recovery efforts. This includes the reconstruction of its schools.

Cyclone Idai and Cyclone Kenneth struck Mozambique in 2019, within six weeks of each other. The country endured back-to-back landfalls at major hurricane force for the first time in recorded history. Idai would have been bad enough—the UN called it "one of the worst weather-related disasters . . . in the southern hemisphere." But Kenneth proved to be the strongest storm ever to make landfall in mainland Africa. International relief efforts gathered only 25 percent of necessary funds during the storms' immediate aftermath. To cover the gap and finance its own recovery, Mozambique was forced to take out millions of dollars of loans from the International Monetary Fund.

These disasters disproportionately harm women, the disabled, low-income, Black, and Indigenous communities, all of which have been marginalized for historical and contemporary reasons. In 2018, when Hurricane Michael tore through Florida and Georgia, it was just the fourth Category 5 hurricane landfall in US history. The areas most affected are some of the poorest parts of the country—impoverished counties of southern Georgia and the Florida Panhandle, which have been scarred by centuries of racism and slavery. Instead of bringing attention to these communities in the aftermath of Michael, most of the media coverage focused on the billions of dollars' worth of ruined fighter planes at Tyndall Air Force Base in Florida.

In Alaska, where 92 percent of the state's revenue is still dependent on the oil and gas industry, summertime now means unusual thunderstorms, relentless wildfires, and unprecedented heat waves. In 2018, Alaska set an ominous milestone: for the first time, the state's annual average temperature crossed 32°F (0°C). On July 4, 2019, as smoke from wildfires obscured the sky, temperatures in Anchorage hit 90°F (32°C), and sea ice near Alaska sunk to a new record low. The permafrost—frozen soil that traps billions of tons of carbon across the Arctic—is melting decades earlier than scientists expected, worsening the effects of climate change and crumbling homes, businesses, and roads—and even entire native communities. A NASA study late in 2019 confirmed that the Arctic had switched to a net emitter of greenhouse gases likely for the first time in tens of thousands of years. July 2019 was the hottest month in recorded history on our planet.

In early September 2019, Hurricane Dorian, another Category 5, stalled over the Abaco Islands in the Bahamas for nearly a day. Despite its destruction, the American press largely ignored Dorian and its aftermath, deciding to cover with great fervor President Trump's use of a black Sharpie to alter an official National Hurricane Center forecast to make it seem as if it were in line with his erroneous tweet stating the storm threatened Alabama instead. But this is how the press often behaves, as if the people enduring the worsening climate emergency are irrelevant, as long as the disasters do not land on US soil.

By every account, Dorian inflicted on the Bahamas the worst single day of weather in the recorded history of the Western Hemisphere: sustained winds of 185 miles per hour, a surge in ocean levels of 23 feet, an unrelenting force that leveled even concrete storm shelters. Thousands of Haitian immigrants, many of whom worked in the luxury resorts on Treasure Cay, lost everything.

"Every morning, you wake up, you open your door and you see the debris and it's just getting to you," Eddie Floyd Bodie, a Bahamian pastor who grew up near where Dorian made landfall, told the *Miami Herald*. "Your mind is wondering what's going on. It's a bad feeling knowing that you used to seeing things that you don't see anymore. What do you say? You say you better try to get adjusted to it, but it's hard. The pressure starts to get to you."

As the year came to a close, a firestorm erupted in Australia on New Year's Eve. In the resort town of Mallacoota thousands of people took shelter on the beach, walled in by rapidly advancing flames on all sides. The fires were the largest in recorded history on the continent, covering an area eighty times the size of New York City. Entire ecosystems were wiped out. In the state of New South Wales alone, an estimated 480 million mammals, birds, and reptiles were killed. Prime Minister Scott Morrison watched fireworks in Sydney Harbor as his country burned.

Climate change doesn't always take on such dramatic forms. More often, it's insidious. Bugs can survive in places they couldn't before, greatly increasing the threat of tropical diseases, even as far north as Alaska and Greenland. In search of cooler weather, trees, birds, mammals, and other species are creeping up mountain slopes and toward the poles. Spring green-up occurs earlier every year, shifting the timing of thousands of species' interactions and rapidly shifting growing zones, which throw entire ecosystems dangerously off-balance. Heat waves have become prolonged and deadlier. Wildfire smoke is aggravating chronic illnesses hundreds of miles away from the flames. Air pollution, worsened by fossil fuel burning, kills more than nineteen thousand people a day, making it one of the leading causes of death in nearly every country on Earth. Young people growing up today are seeking treatment for mental health issues in numbers never seen before, in part because they are not always sure they'll have a livable future.

It can't go on like this. Somehow, some way, we have to learn how to care about one another again.

When we read stories about how climate change is altering the world, journalists often focus our attention on people and places far removed from our everyday experiences. Polar bears are majestic and fascinating creatures, but virtually none of us will ever interact with one. For the millions of people who do live in the Arctic, mass starvation of other animals is happening with increasing regularity and creating a much more immediate impact on their lives. In recent years, about a quarter of the caribou in Russia died due to unseasonably warm winter weather, which transformed the normally soft snow into a sheet of ice, preventing them from accessing the grass below. The loss of sea ice doesn't affect only polar bears, which depend on the ice to hunt; it's killing off the region's entire food chain, from migratory whales to plankton. Sea birds, like puffins—leading indicators of oceanic health—are also experiencing a rapid demise. Just inland along the Arctic coastline, the growing season has almost doubled in length over the last decade as the open water offshore has transformed the landscape from tundra to a humid shrubland. The environment has been thrown into a tailspin.

Offshore, the opening of new waterways has transformed Arctic fishing industries: In Greenland, mackerel—migratory fish that also live in tropical waters—had never been seen locally until the start of the twenty-first century. They now arrive there every year, making up one-quarter of Greenland's fishing economy. Salmon, too, pushed to the point of local extinction in California, have occasionally been spotted in the Arctic. All these changes are happening while the people who have lived there for thousands of years fight to preserve their ways of life and fend off greedy companies looking to establish Arctic shipping lanes and claim mineral rights.

Meanwhile, the rest of us are experiencing our own surreal encounters with the rapidly transforming planet, every day. In 2016, an octopus splayed out inside a Miami Beach parking garage went viral on the internet. Climate reporter Brian Kahn has kept a running log of these Dali-like moments on Twitter, labeling them "Postcards from the Anthropocene." Recent features include: a man fly-fishing next to the Washington Monument (who ended up catching a carp); two men playing golf while a raging wildfire burns in the background; a man mowing his lawn while a tornado churns near his backyard; sunbathers during a heat wave in northern Finland sharing the water with a caribou; a police boat cruising down the center of a flooded interstate in North Carolina; servers and patrons carrying on with their dinner in ankle-deep water at a restaurant in Italy; and a firebomber airplane scooping up water just offshore from a surfing beach in California.

That Miami Beach octopus became famous because of a king tide, a phenomenon that occurs during the twice-monthly gravitational alignment of the Earth, the sun, and the moon, and is exacerbated by the rise in sea levels. It's this kind of gradually escalating flooding that will likely force Floridians to permanently retreat from the coastline, not a devastating hurricane. By the 2040s, within the time span of most homeowners' mortgages, the Union of Concerned Scientists predicts that chronic coastal flooding—defined as flooding that happens twenty-six or more times a year—could envelop 300,000 coastal homes currently worth a combined total of more than $100 billion. And that's just in the United States. By the end of the century, that figure could balloon to hundreds of trillions of dollars worldwide in the worst-case sea-level-rise scenarios, a truly daunting prospect that climate scientist James Hansen has characterized as "the loss of all coastal cities, most of the world's largest cities, and all of their history."

Changes like these, at once pervasive and insidious, define our new planetary era. The simultaneous disturbance of nearly every ecosystem on Earth is helping opportunistic, weedy species flourish while specialized plants and animals race to adapt. There's no better example of this than the worldwide boom in jellyfish, which have become a danger to swimming and power plants, and are newly dominating ecosystems all over the planet. Jellyfish expert Lisa-ann Gershwin said that after at least 585 million years of existence, the current moment might be the best thing that's ever happened to jellyfish. "Imagine a world where somebody else warms up the water that you're living in so you grow faster, eat more—of course there's more food available because the competitors are gone . . . you reproduce more and you live longer to do more of it," she explained. "You would be pretty happy."

The same thing that's happening in the ocean is happening on land. Warming temperatures are increasing the metabolism rate of soil microbes like bacteria and fungi. As the tiny organisms break down decaying plant and animal matter at a faster rate, they release more highly potent greenhouse gases, like methane, which escalate global warming. Since microbial activity underlies literally all life on Earth, the fact that the behavior patterns of these organisms are rapidly changing is worrying.

Of course, there are more visible changes on land too. Many of the plants and animals that dazzle us, that we sing songs and write poetry about, are being profoundly disturbed. Fireflies, a staple of childhood summer evenings, are shifting to earlier emergence as springtime warms and rainfall patterns become more erratic. Rhododendrons, whose flowers are turned into juice in the Himalayas, now bloom an astonishing three months earlier in India. The iconic bluebell woods in Britain may not survive because the fragile flowers aren't resilient enough to keep up with the shifting seasons. At the current rate of warming, the Douglas-fir, a favorite for Christmas trees and the anchor of Pacific Northwest forests, may disappear entirely from the coastlines there by the end of the century.

In nearly every culture, in nearly every country around the

world, people have always kept closely attentive to subtle changes in their surroundings and relied on input from the natural world to make sense of the orderly passage of time. Using records of everything from birdsong to flowers, there are now more than 26,500 independent signs of climate change. For more than a thousand years, the people of Kyoto, Japan, have kept track of the date of the appearance of cherry blossoms. The pink and white blooms now appear in April on average ten to fifteen days earlier than they did in the past millennium. In Micronesia, the greatest sailors who ever lived navigated thousands of miles over open ocean by watching the stars, the waves, the clouds, and the birds. Their descendants are now watching those same waves encroach on their homelands with increasing alarm. In Australia, more than forty thousand years of weather watching has long guided Aboriginal understanding of the seasons, and a ten-thousand-year oral history tradition is now helping them prepare for climate change by recalling and learning from a long-standing connection to the land. In the Okavango Delta region of Botswana, one of the most highly variable climates on Earth, traditional methods of weather forecasting have served as successful strategies to anticipate the future for generations.

But a new era of urgently paying attention to nature has arrived. Phenology, the study of animal and plant responses to the weather, has in the past decade gained renewed importance in the context of rapid global change. Traditionally, phenology has focused on seasonal fluctuations of plants and migrating animals. But lately, the behavior scientists are seeing is so unusual that it's starting to seem more like tipping points are here.

One thing is clear: in recent years, these trends are only getting worse. We no longer have just freak weather; we have freak seasons. In 2012 and 2017, across most of North America, spring arrived exceptionally early, a trend that may soon become a nightmare for agriculture. In March 2012, overnight low temperatures broke Michigan's previous record high temperatures in several cities, throwing the state's apple and cherry harvests into a tailspin. It was full-on summer weather—with temperatures as warm as 90°F (32°C)—in a place that usually sees snow on the ground until April. In 2017, another early spring, on the East Coast, threw off bird migrations and led to a mismatch between food availability and the arrival of the hungry animals from far away. It was also one of the worst and longest-lasting allergy seasons in memory.

Theresa Crimmins, associate director of the USA National Phenology Network in Tucson, Arizona, chronicles these kinds of changes as part of her life's work. What she is seeing shocks her. "If we don't take urgent action," she told me, "climate change projections suggest that years like 2012 and 2017 in the US could become normal by midcentury."

Around the world, phenologists like Crimmins watch in real time, mouths agape, as the data pour in from thousands of parks, gardens, museums, and citizen scientists. When the change is this rapid, sometimes adaptation simply isn't possible.

In moments like these, people can lapse into solastalgia, a deep longing for the natural world that we know is never coming back. Recent studies have found that our brains keep only a two- to eight-year record of weather extremes. In a world that is rapidly changing, we literally have no frame of reference for how unusual our climate-related experiences are. Climate change is changing who we are, changing our sense of place, and loosening our grasp on reality. It's no wonder we sometimes feel like we're losing control.

* * *

My own story, of course, also has climate connections. I grew up in Kansas. Land there has already been utterly transformed from what the first European colonizers called the "Great American Desert," even before climate change. Kansas had been home to the Kaw people for thousands of years and is now home to some of the most productive farmland in the world. But in a warming world, industrialized agriculture imbues the region with massive risk. Western Kansas, for instance, gets about the same amount of rainfall each year as southern Arizona. For decades Kansas farmers have increasingly made use of groundwater pumped from the Ogallala Aquifer to grow corn. What will happen when that aquifer inevitably dwindles to the point it's no longer useful for agriculture? What will happen to my hometown? What will happen to the countless other small towns and the families who have called them home for generations?

During the spring of 2019, the whiplash effect of climate change was on display on a massive scale: just months after widespread drought in 2018 ruined fields in Kansas and Missouri came the wettest year in US history. Rivers spilled over to flood dozens of towns from the Dakotas to Louisiana, effectively turning them into islands. At one point, between Omaha and Kansas City, every levee was breached on the Missouri River. An aerial view at the Nebraska state line visually confirmed that the overflowing Missouri looked more like an ocean than a river. Cities from Saint Louis to Baton Rouge experienced the longest-duration flood in history. And as the spring flood stretched into hurricane season, a new type of compound flood threat emerged during July's Hurricane Barry: for the first time in history, New Orleans completely sealed itself off and waited out a threefold attack from rain, river, and rising seas.

By midcentury, on our current course, long-lasting droughts will have become more common and more severe across the western part of North America, including in my home state of Kansas. Recent studies predict that, later this century, a megadrought will persist for decades. Indeed, it may have already started. A drought like that would create a new normal. Even if the amount of rainfall doesn't change, rising temperatures will boost the speed of evaporation and make the rain that does fall less useful to plants, altering every aspect of life in the place where I grew up.

Chronicling the human-induced destruction of the natural world in real time isn't a career I'd recommend for people who get emotionally attached to their work. Watching all this happen with the keen eye of a climate reporter and the broken-open heart of a parent is difficult to handle. By the time my two preschoolers, Roscoe and Zeke, are as old as I am now, the world will be transformed. I'm not quite sure what to do with that information. My job is to study the weather and climate on a global scale, and I'm not even able to fully comprehend how radically things in our neighborhood will change in my kids' lifetimes.

Our current home in Saint Paul, Minnesota, sits a half mile from the Mississippi River. Outside of Alaska, Minnesota is the fastest warming state in the country. We are losing populations of native species of plants and animals at an alarming rate, just as fast as in the Pacific Islands. And just as it is in Puerto Rico, the weather is becoming more erratic and extreme. At another time in history, Saint Paul would be the perfect place to raise a family. Now, though, there are days when a deep anxiety takes over and I'm paralyzed with worry about what will happen if we don't radically change course.

I often wonder:

Will the forest in our backyard survive?

Will our mosquitoes begin to carry tropical diseases?

Will a flash flood scour away our street?

Will our neighbors continue to welcome refugees and climate migrants, or will they turn them away?

Will we come together and create the change we need soon enough?

For those of us trying to make sense of the current moment, the fact that the weather is permanently changing might already feel obvious. But planetary-scale changes aren't supposed to happen in a century—let alone in the amount of time it takes for my boys to grow up. The more I learn about what's happening, the harder it is for me to write about climate change from a dispassionate and objective point of view. I cannot ignore the fact that I'm a real person raising kids who will be here for (hopefully) many more decades. In a few years, they'll start asking tough questions. I want to have answers. That makes climate change personal. For me, that makes it about love.

Researching and writing this book has been a years-long emotional struggle that started when my then-wife and I were still expecting Roscoe's birth. It has continued through multiple moves across the country, a divorce, mental illness, and starting over several times. And yet I feel as if I'm one of the lucky ones. I understand that climate change is not about me or my feelings. Still, I know there is nothing else I can justify doing at a time when so much is at stake. Most of the time, I feel like I can't keep up. If I'm feeling like this, from my privileged position as a white male in an affluent and stable country, it's impossible to overstate the courage and the sheer will of survival of people in places like the Abacos. It's no wonder that young people all over the world are feeling an outpouring of climate anxiety and pre-traumatic stress syndrome. They don't know if they will have a future at all.

The events of the past few years show that we've passed the first major climate tipping point. This carries deeply profound implications for us all, even if we don't yet realize it. In a matter of months, climate change has moved from something that was supposed to happen in the future to something that is causing immediate and irreversible damage across the planet.

So what does it mean for us, as individuals, and for our society as a whole? How do we process the overwhelming and terrifying knowledge that we are living during such a rapidly changing era of Earth's history? And more important, why aren't more people freaking out about this?

We've all experienced profound loss in our lives—a bad breakup, incurable diseases, tragedies that feel like the world is crumbling in on top of us. What might it mean for an entire country or society or civilization to walk together, hand in hand, through stages of grief and loss and depression and mourning, at the same time? What would it be like to anticipate not only our own death but The End, the apocalypse? That's what it's like to be alive in the world these days.

Beyond the forthcoming technological advances that will occur during this climate emergency is a revolution in human psychology—the way we view ourselves and our place in the grand order of things. Rising seas, mass migrations, and escalating extreme weather events mean the idea of humanity's dominion over the natural world is about to get turned on its head. "If we don't demand radical change," activist and author Naomi Klein said, "we are headed for a whole world of people searching for a home that no longer exists."

Individually, each of us will have to go through a grieving process for the loss of a world we believed in our bones would always be there. Collectively, to help mourn and accept this loss, we will have to share with one another alternative visions of a shared future, stories about how climate doom is not inevitable, and what the future Earth might look like if we do what is required—and still entirely possible—to hold off the greatest threat to our very existence.

What you're feeling right now is not unique to you. And just because others might not be feeling it yet doesn't diminish the intensity of your experience. You are not alone in bearing this existential dread, this fear of the future, this hopelessness. But just as important, this feeling does not have to define you. In fact, your energy, your emotion, your desire to right this wrong, is a critical part of the solution. It is precisely because you care and want things to change that you feel this way.

In January 2017 I started counseling for climate-related anxiety. To this day, I struggle with how to focus my attention and energy to fight climate change. But I can tell you that part of the answer is just talking about it, and to know you are not alone. Drawing from the science of psychotherapy, I believe it is possible to spontaneously and emergently solve problems in our lives—and in society—simply by talking about them. The physical act of speaking changes the way your brain works and causes you to think differently. And if we ever needed to think differently, it's right now.

In the hundreds of conversations I had while researching this book, I kept coming back to one inescapable conclusion: as long as we are still here, it means we haven't yet lost the fight. And that realization gives me a glimmer of hope.

I've come to accept the fact that we're entering a scary time of profound change. I'm not going to push for any specific radical lifestyle changes or for any revolutionary new political campaigns in response. That's not my place. Instead, I'm going to encourage all of us to explore possible futures based on the latest science and continue to have faith that the conversations themselves could be transformative. It's likely that you—yes, you—have an idea that I've never thought of. It could be your idea that winds up making all the difference. Our time requires us to listen to the people who have been ignored and unheard because what they have to say is inconvenient to the people in power. This problem affects all of us, and so the future will require the creativity of all of us.

The only thing I've become convinced of is that we need a revolution, which you can define however you want. No matter how we define it, though, nothing less than a complete transformation of almost everything we know will be enough to get us through the next few decades. The old world is dead—what comes next is up to us.

I'm a journalist. I can't afford solar panels for my house or a Tesla. I often feel uncomfortable about joining a protest. But I can talk with my friends about how I feel. I can imagine what I hope to see in the future. And then, just maybe, we can brainstorm things that we can do together to build upon our shared desires for justice. Maybe, together, we can build a movement focused on love and repair.

The biggest climate lie is that individual action is the only answer—that's a recipe for burnout and continued disaster. Individual action is only useful when it helps bend society toward radical change. And the only way to create lasting change is to work toward a future in which everyone matters.

* * *

At my family's farm in northeast Kansas, a watershed creek meanders through the property. Last winter, during a particularly cold stretch, Roscoe and I walked out on the frozen creek with my dad to visit a beaver den and check out the hole in the ice where the animals remain active throughout the winter, the kind of thing that people here have been doing for thousands of years. It may just have been the juxtaposition of watching my oldest child soak up some of the lifetime of knowledge my father has gained in this corner of Kansas, or maybe it was watching the plants and animals slowly adapt to the changing weather, now that thick ice is rare in the wintertime. The afternoon felt like a relic from a forgotten time—an eerie and tangible reminder that what we knew and loved in the past will never be possible again.

This is what climate change feels like, whether you live in Kansas or Puerto Rico, Fiji or Houston: persistent chaos, a subtle unknowing, a sudden loss of the familiar, the reshaping and redirecting of dreams, an unlearning of expectations.

We can no longer deny that weather in every corner of the Earth is different now. That change is because of us. And we have the power to choose a different path.

In the waning days of 2018, a special report from the Intergovernmental Panel on Climate Change sent shockwaves through the world. The group of scientists tasked by the United Nations with compiling the report found that, absent heroic effort, the world has locked in changes to the atmosphere and oceans so dangerous that they could pose an existential risk to civilization. That stark assessment came, after years of deliberation, at the request of the world's most vulnerable nations to study how to limit global warming to no more than 1.5 degrees Celsius (2.7 degrees F), which scientists now affirm would give us the best chance of preserving the stability of the world's interconnected ecological and societal systems. If no action is taken, the world will reach 1.5 degrees as soon as 2030, jeopardizing the lives of several hundred million people. The report was firm on this point: to ensure a livable future, we have to do everything we can, right away. There is no way around this anymore. Existential risks greatly escalate if the world lets the 1.5 degrees goal slide to 2 degrees, much less the 3.4 degrees we're currently on pace for. Trying to decide between 1.5 degrees and 2 degrees is like choosing between *The Hunger Games* and *Mad Max*, as Kendra Pierre-Louis, a climate reporter for the *New York Times*, put it.

One of the report's lead authors, German climate scientist Hans-Otto Pörtner, told me the IPCC's assessment was a milestone, a dire warning. "If action is not taken," he warned, "it will take the planet into an unprecedented climate future if we compare it to what has happened in all of human evolutionary history. . . . Climate change is shaping the future of our civilization."

Just as the weight of the IPCC report started to set in, another arguably worse bombshell landed. A team of scientists from the UN-sponsored Intergovernmental Science-Policy Platform on Biodiversity and Ecosystem Services (IPBES) found that we could be setting in motion a mass extinction with little parallel in the history of the Earth. Without radical change, one million species are at risk of extinction in the next few decades, one in eight species on the planet.

That day, *The Guardian*'s headline summed up the stakes: "Human Society Under Urgent Threat from Loss of Earth's Natural Life." What was equally clear in the report, but not necessarily in the news headlines, is that the scientists pinpointed the extractive, exploitative nature of the global economy as the root cause. The IPBES report found that current methods of agriculture, deforestation, and urbanization are crashing the planet's ecology at a rate tens to hundreds of times faster than at any point in the past 10 million years.

What we're doing to the planet's climate now is about 170 times faster than natural forces, according to one recent study. Some of the most rapid climate change in Earth's history occurred about 250 million years ago, during a period known as the "Great Dying." About 90 percent of all species on Earth disappeared, the worst mass extinction event in Earth's history. During that period, large volcanic eruptions increased the atmosphere's greenhouse gas content, leading to an abrupt planetary warming over a period of about twenty thousand years. Now a change of similar severity could happen in just a century or two. Will Steffen, an Australia-based climate scientist who helped conduct the IPBES study, said that when viewed through the lens of geologic history, "the human magnitude of climate change looks more like a meteorite strike than a gradual change."

Gretta Pecl, director of the Centre for Marine Socioecology at the University of Tasmania in Australia, said this fact changes everything. Species that are migrating due to climate change don't respect the borders of nations or national parks. "The old ways of conservation are entirely out the window," she told me. "We used to want to protect things in little particular areas, but things are not the same and never will be the same. We have to actually have some kind of large-scale strategic way of thinking about how to manage species."

Of all the ecological changes described in the IPBES report, the most devastating so far has been the loss of a quarter of the world's coral reefs, which was accelerated by the El Niño of 2015, a record-breaking event that shifted weather patterns around the world and triggered permanent changes to maritime and terrestrial ecosystems everywhere.

El Niño is a natural, periodic warming of the tropical Pacific Ocean, but the 2015 El Niño was anything but normal. Recent studies show that the warming oceans could be altering the El Niño cycle, making strong episodes more frequent. And all this warm water is really doing a number on the oceans, the basis of the planetary food chain. The loss of the coral reefs means we could already be entering an era of mass extinction with entire ecosystems being snuffed out almost overnight. And if we lose the reefs, we lose one of the anchors of life on Earth.

"This is a huge, looming planetary crisis, and we are sticking our heads in the sand about it," Justin Marshall, a coral researcher at the University of Queensland, told the *New York Times*.

Coral reefs are more important than many people realize: taking up just 0.2 percent of the ocean, they support about a quarter of all marine species and provide support to livelihoods of 500 million people. But beyond that, healthy reefs are stunningly beautiful. They're part of what makes life on Earth so special.

The largest coral atoll in the world, Kiritimati (a Kiribati spelling of the English word "Christmas"), was, up until recently, one of the most pristine marine ecosystems on Earth. Over a period of months, a team of researchers—led by Julia Baum, a biologist at the University of Victoria, and Kim Cobb, a climate scientist at Georgia Tech—took comprehensive measurements of the reef's health, or lack thereof in this case. Their assessment was shocking: about 80 percent of the coral colonies at Kiritimati were dead, and another 15 percent were severely bleached and likely to die. It's as if someone decided to cut down 90 percent of all the giant redwoods in Sequoia National Park over the weekend. In Kiritimati, an entire ecosystem has essentially blinked out of existence.

I spoke with the team by satellite phone on one of their last days of dives, and the shock in their voices was palpable. "There's a good chance that this reef will never be the same," said Cobb, fighting back tears. "It's a wake-up call."

Cobb has been working at Kiritimati for years, so she's deeply familiar with what it's supposed to look like. From cores that Cobb's team has analyzed, she estimated there's been nothing like the current die-off in Kiritimati in the seven thousand years of ancient coral history there. Global warming will make the pressure on corals even worse in the coming decades, and many of the world's reefs can expect future bleaching events to occur more frequently. For some, like people in Kiritimati, the worst global coral bleaching episode in history may have already been a point of no return.

"This is a story that the rest of the world should hear," Cobb said, and added that the death of the corals at Kiritimati is proof that "climate change isn't just a steady linear progression to some different kind of planetary state. . . . Climate change will occur through these kinds of extremes. It's like a staircase to a different kind of system."

After decades of sharply rising emissions, it's clear that the Earth's climate system is stair-stepping into the unknown. Scientists have identified dozens of tipping points and nonlinear properties of the climate system, which simultaneously provoke fear and hope. Fear for obvious reasons—the relative stability we have had until very recently could quickly unravel—and hope because no matter how bad things *could* get, quick action will almost certainly be enough to prevent the worst-case scenario. The most important implication of our nonlinear climate system is that a climate apocalypse is not inevitable. At the same time, every moment we wait locks us further and further into irreversible ecological change.

Some of the media coverage of the IPCC report framed its stark conclusions in the too-familiar phrase of a time limit for action. "The World Has Just over a Decade to Get Climate Change Under Control," read *The Washington Post*'s headline. But that's absolutely the wrong way to frame this. We have only a decade left to *finish* our initial coordinated retooling of society to tackle this challenge. The scientists were quite clear about this. By 2030, we'll need to have already cut global emissions in half (45 percent below 2010 levels), which according to the IPCC would require "rapid, far-reaching, and unprecedented changes in all aspects of society."

Those transitions must begin immediately. There's no time to waste.

"All options need to be exercised," said Scottish climate scientist Jim Skea, one of the report's lead authors, at the report's press conference. "We can make choices about how much of each option we use . . . but the idea you can leave anything out is impossible."

That radical change needs to focus on not only preventing apocalypse but also building up a picture of a future that's worth fighting for. Imagining the world as a burning hellscape is, for some reason, much easier than imagining a world where we come together and build a new version of human society that works for every person and every species. We've got to do the hard work of fashioning a culture of radical, unrepentant, courageous hope.

Kim Cobb has spent her entire career studying the coral reefs of the central Pacific, but the most recent bleaching event clearly changed her. Since her phone call with me from Kiritimati, she has become an activist. She's organized protest rallies. She makes presentations to youth organizations about ocean science. She bikes to work. She's vastly reduced her use of cars and airplanes. She's testified to Congress about the need for radical action. She's transformed her life. Still, she is filled with doubt, and setbacks are numerous. After the IPBES report came out in 2019, Cobb told me: "Sometimes, when I get out of bed in the morning, I really wonder why the hell it all matters."

I find inspiration and peace in her courage. Changing the

world is hard, and she knows there's no time to spare. Still, she remains hopeful and courageous in the face of the seemingly impossible. "One of the very few benefits of this new struggle," she told me, "is [learning] how to work together and take absolutely nothing for granted."

## BECOMING RADICALS

Radical change is now inevitable. It's up to us to foster new conversations that confront this truth head-on yet are imbued with the kind of patience and care necessary to finally turn our intentions into meaningful action. We must begin a transition to a new kind of environmentalism that reflects the way the actual environment works. Not a demonstration of individualism or moral superiority, but an actionable, scalable model for a new way of life rooted in collective support and universal justice.

Throughout the next decade, we will experience both creative imagination and creative destruction, which is likely to produce a deep and abiding sense of civilizational anxiety. Certain times and certain spaces make us feel uneasy by their very nature: dark, empty stairwells; rest stops along the highway; the aftermath of a breakup. These are liminal spaces. If you find yourself spending more than several minutes in a liminal, or transitional, space, your inner lizard brain wants to flee—alarm bells begin to ring, something isn't right. Because they transcend our normal understanding of how the world is supposed to work, they feel haunted. Liminal spaces are temporary, incomplete, and portentous. They imply possibility to such a degree that it is sometimes literally frightening.

Right now, Earth—the entire planet—is a liminal space. We are starting to learn this fact, that radical change is inevitable, yet we don't know exactly what it's going to look like. But this time is also vitally necessary in order for us, as a species, to figure out what we're about to become.

Samantha Earle is a philosopher at the University of East Anglia who specializes in liminal spaces. She firmly believes that Western civilization is currently ripe for radical change. Talking with her was the most transformative conversation I had while putting together this book. Philosophers like to speak of an "imaginary," or a guiding framework, for organizing society. Earle isn't under any illusions that this process is going to be easy. But just as climate scientists are certain that our current path will lead to inevitable destruction of our planet's fundamental life-support systems, Earle is certain that our prevailing guiding frameworks for civilization cannot last.

"We're at that time where the problems of the world just can't be answered by the prevailing imaginary," she told me. "We are in a time of breakdown."

That this breakdown in our society is happening precisely during a breakdown of our atmosphere and ecosystems is no coincidence. Through our daily thoughts and actions, we constantly reproduce our society. Most of that happens unconsciously or by routine or habit, or because exposing ourselves to new thoughts and actions involves risks we're currently unwilling to take. But everything we do every single day is a choice, which means different choices are possible.

"The major problem with society is that we don't even recognize that we have a particular imaginary, that this is not how things have to be," Earle told me. During normal times, "we lack critical awareness, and we lack the capacity to radically imagine."

This is why opponents of radical change have successfully categorized efforts to protect the environment or reverse the effects of climate change as personal sacrifices. We focus on the inconvenient downsides of taking the train over a car-share service, for instance, instead of imagining or working toward a world where driving in cars simply doesn't exist because cities are designed with people in mind. We choose to keep funding fossil fuel companies because they've spent billions of dollars making sure our lives appear easier when we buy their products. The status quo is comfortable for a reason: it makes daily life easier to manage, especially when the alternative doesn't yet exist—or, more accurately, when those in power are actively opposed to making a better world a reality.

In the midst of a liminal space, however, anything is possible. "Liminal space is a time of radical uncertainty where the foundational concepts of the way in which we've been living, and around which society is organized, no longer make sense," Earle told me. "It's not just that we're unable to make sense of the problems that we're facing. We can't even conceptualize them. It's a time almost of being suspended—it has profound existential implications. You still have an imaginary, only it doesn't work, and you don't have another one in place yet, and everything is up for grabs."

The first step in moving through this space is to acknowledge the discomfort we are in right now as individuals and as a society. There's creative power in that discomfort—it helps us to see the possible paths before us with clear eyes. It helps us to imagine something better.

Radical change isn't something that most of us choose to think about very often in our free time. It's unpleasant to consider how quickly our planet is shifting into a new and more dangerous version of itself. We don't want to contemplate the delicate balance between environmental calamity and our drive to make our lives better, but we don't have that choice anymore.

So where do we go from here? How do you and I use this existential dread as a motivating force to imagine a different path? And then how do we build that new path?

"Climate change is first and foremost a problem of our relationship with the future," author and philosopher Alex Steffen told me in the early stages of writing this book. "The way that we think about the future is almost entirely cultural. If we can't adapt our cultural perspectives to include the idea that we need to be in a sustainable relationship with long-term systems, none of the other actions we need to take are going to happen."

A recent study mapped out a blueprint for how to enact the kind of change Steffen is talking about. "It's way more than adding solar or wind," said Johan Rockström, the study's lead author. "It's rapid decarbonization, plus a revolution in food production, plus a sustainability revolution, *plus* a massive engineering scale-up [for carbon removal]." According to Rockström and his team, to get the math right, we need to cut global emissions in half by 2030, then cut 2030 levels in half by 2040, then cut 2040 levels in half by 2050. Even these heroic efforts will secure only a roughly 50-50 chance at achieving the 1.5-degree limit in temperature rise the world agreed to in Paris. By the time all that is done, my boys will be only a few years older than I am today. Zeke won't even be able to vote until 2034. By then, if we do nothing, we'll have lost our only remaining chance at a livable world.

How this will actually play out is, of course, unknowable. But we can make some educated guesses. Meteorologists like me are used to predicting the future in a way most other professionals don't. We can see the range of possible future weather events and envision what they will be like. To date, we haven't done a good job imagining how the weather years from now will affect society. But for the first time, the Rockström study provides a simple way to think about it: transformational change, decade by decade, for the next thirty years.

Very few people are already living in this future world. Even fewer understand that our current actions require systemic change. And none of them are as famous, or as influential, as Greta Thunberg. The improbable fact that Greta, a seventeen-year-old Swedish girl, is one of the most powerful people on the planet is further evidence that the old rules no longer apply. Her ability to speak disarming truths about our climate emergency goes far beyond her commitment to a zero-carbon lifestyle. Her power rests in the fact that, when it comes down to it, she knows that these actions are only meaningful if they create radical and systemic change. The fact that her actions are so simple means that everyone can join her. She has shown us a way out of this liminal space.

A recent survey revealed that one in five people in the Western world say they are now flying less because of the movement Greta helped start. In Sweden, there's even a new word—*flygskam*, "flight shame"—and the airline industry there, built on the promise of eternal economic growth, is starting to get scared. SAS, the Swedish national airline, has called it the "Greta effect." Representatives from all twenty-nine companies on Sweden's benchmark stock index now say that air travel in the country is likely to have peaked, permanently, because people are starting to understand that, in a climate emergency, there is no feasible alternative for jet fuel in the near term. And batteries weigh too much for jet-based air travel as we know it—that's just simple physics.

The first glimpse I got of Greta was late at night, a silhouette wearing a blue jacket, standing in my living room, reading a handwritten note my toddlers and I had made for her. She and her father, Svante, had arrived at my home in Minnesota from Iowa City for a few hours of rest, in a Tesla loaned to them by Arnold Schwarzenegger. Over the past few months she had addressed the European Parliament in Brussels, boarded a boat in the United Kingdom and sailed across the Atlantic, given a speech to world leaders at the United Nations General Assembly in New York, presided over the largest climate rally in world history in Montreal, and met a famous vegan icon pig in Ontario. The next afternoon she would depart for Pine Ridge and Standing Rock in South Dakota to meet the people who inspired her to take this trip in the first place. She was, in a word, tired.

During our time together, we were able to have breakfast and go for a walk—normal activities for two adults, two toddlers, and a teenager. A soft early autumn rain fell on us. Roscoe fussed that we were rolling him around in our red wagon instead of letting him ride his bike. Other than that, the day was like any other. Greta and her dad felt just like friends visiting for the weekend.

On our walk, Greta paraphrased the points from her recent speeches: we are still in the denial phase, she told me, still trying to convince ourselves that, among other things, we are doing enough, or that the problem isn't as bad as it is, or that we don't need to act like it's an emergency. I've heard people make the same point Greta was making, but hearing these words directly from her, in my own neighborhood, with my youngest child in a wagon and my oldest child in my arms, at a moment when everything I love felt so delicate and impermanent that it all could go away in an instant, hit me square in the chest.

Just weeks after starting her school strike outside the Swedish Parliament in Stockholm in August 2018, then-fifteen-year-old Greta Thunberg addressed world leaders at the UN climate summit in Poland. She spoke with the moral clarity of an entire generation sentenced to a desperate future:

> *Our civilization is being sacrificed for the opportunity of a very small number of people to continue making enormous amounts of money.*

*Our biosphere is being sacrificed so that rich people in countries like mine can live in luxury. It is the sufferings of the many which pay for the luxuries of the few.*

*The year 2078, I will celebrate my seventy-fifth birthday. If I have children maybe they will spend that day with me. Maybe they will ask me about you. Maybe they will ask why you didn't do anything while there still was time to act.*

*You say you love your children above all else, and yet you are stealing their future in front of their very eyes.*

*Until you start focusing on what needs to be done rather than what is politically possible, there is no hope. We cannot solve a crisis without treating it as a crisis.*

*We need to keep the fossil fuels in the ground, and we need to focus on equity. And if solutions within the system are so impossible to find, maybe we should change the system itself.*

*We have not come here to beg world leaders to care. You have ignored us in the past, and you will ignore us again. We have run out of excuses, and we are running out of time. We have come here to let you know that change is coming, whether you like it or not. The real power belongs to the people.*

With these words, Greta Thunberg kicked off a worldwide revolution.

Since this speech, tens of millions of people of all ages in 183 countries around the world have joined her climate strike, buoyed by her clear vision of a world that's not just survivable but thriving. People want something to hope for. People want to be asked to do something big. People want a chance to be courageous, a chance at shaping their own future. That is what this moment provides.

Of course, this remarkable moment in history didn't just happen spontaneously. For centuries, millions of people, led mostly by women of color, have laid this groundwork. "I am not saying anything new," Thunberg wrote in a Facebook post a few months into her strike, which was inspired by youth protests in the United States on climate change and gun violence. "I am just saying what scientists have repeatedly said for decades." On the sidelines of the official United Nations climate negotiations each year, the voices of children in the global south, where forests continue to be devastated and extreme weather threatens normal existence, have long been drowned out. Greta's message is breaking through and resonating today, I believe, because the environmental emergency that rich countries have inflicted on poorer countries for centuries has finally come home to roost.

We know that younger generations should be able to feel confident they're inheriting a world that's survivable. As that future has been put up for grabs, Greta has been able to amplify the kind of language that has been in use for decades among those in the environmental justice community where the act of producing and using fossil fuels has annihilated places, either through mountaintop coal mining, cancer alleys, or toxic groundwater, places like the forests and rivers of Appalachia and Amazonia, the coal export terminals of Australia, and the neighborhoods of Flint, Michigan. The message from Greta is as clear now as it has always been: Our future is not up for debate. We all deserve to survive and to thrive.

Greta's rise to the de facto leader of a global youth movement is evidence that things are different now. We've known about climate change for a long time, and we've known the solutions for a long time: phase out fossil fuels, transform agriculture, restructure the way we live, and move. The solutions are the same as they always were. But something important about this moment is different: we're finally ready to listen.

During our walk in the rain, I asked Greta what she thought the system would look like once we change everything.

"I don't know," she said, "it hasn't been invented yet."

It takes courage to try to change the world. It takes even more courage to admit we don't yet know how we can change it. The kind of attitude Greta exudes in her public work and, in private, with me and my family—the courage of both knowing and not knowing simultaneously—is the foundation of the work ahead of us and a prescription for how to accomplish it: an openness to a world beyond our ability to anticipate.

Greta is a regular person; she doesn't act like a celebrity. What makes her special is that she has read and understood climate science and is living her life as if the facts matter. The science, on its own, is revolutionary enough. What she's doing is clearly very tough on her. The week she visited my home she very nearly won the Nobel Peace Prize—it's hard to imagine the kind of pressure the world has put on her shoulders. Hers is a unique expression from a unique person at a unique moment in history. Over and over she repeats that this movement for climate justice isn't about her. It's about everybody. The children on strike are not trying to inspire you. Or even impress you. They shouldn't even have to do anything. The grown-ups, the literal adults in the room, should be acting on their behalf. Her actions are not divisive; they make perfect sense. It is the world's adults, by denying the truth of what this moment requires of us, who are acting divisively.

Like no one before her, Greta has galvanized the world's attention on the most important problem in human history. For this, she is regarded as both a hero and a villain, depending on your willingness to accept the blunt truths she tells. In a more normal time, world leaders would have taken the advice of scientists decades ago, when they first warned that the entire basis of powering our global economy was quickly leading us toward mass extinction. In a more normal time, a generation of young people around the world would not be experiencing "pre-traumatic stress disorder," wondering if they would have a future at all. In a more normal time, Greta would have never set foot in my living room or played with my two boys.

After she and her father left, I realized her beautiful humanity is what makes her so dangerous to the status quo. I recognize this same quality in so many people in my professional and personal life. At times, I recognize it in myself. If we all share the same revolutionary humanity as Greta, then what's our excuse for not embracing it?

* * *

On the day the IPCC special report came out in 2018, I committed myself to courage: the courage to tell the truth about the biggest problem our civilization has ever faced, and to help other people to do the same. There is no time left for restraint or to ask for permission. The basics of climate science are easy. We know it is entirely human caused. Which means its solutions will be similarly human led. And since time is running out, those leaders are going to be you and me.

In 1988, thirty years before this IPCC report, NASA climate scientist James Hansen, in his testimony to Congress, was the first to announce to the general public that climate change had arrived. That moment is, suddenly, a long time ago. In all those years since, we've watched as his predictions have come true, and we've done essentially nothing meaningful about it.

In 1992, nearly every nation on Earth signed the United Nations Framework Convention on Climate Change, an initial pledge to tackle global warming. The treaty was designed to coordinate global action to avoid "dangerous anthropogenic interference with the climate system." Scientists have spent the subsequent decades trying to examine which tipping points would be reached first, and how much might constitute a "dangerous" amount of overall warming—generally agreed to be the point after which humans might begin to lose the ability to rein in the impacts, leading to permanent and potentially catastrophic changes to our planetary life-support system. The findings of the IPCC and the IPBES prove that we're now into the heart of an era of "dangerous anthropogenic interference with the climate system," and we're still making it worse.

One of the most shocking truths of our climate emergency is that half of all of humanity's carbon emissions since the dawn of the Industrial Revolution has been released since 1992. Our planet's carbon emissions in 2019 were the highest in human history. What we've done has not been by accident. We've continued warming the planet with full knowledge of the science. We knew that the changes would linger for millennia, borne by people who did the least to cause them, in order to make the richest people who have ever lived even richer. Climate change has been a choice, and our leaders have again and again chosen to do the wrong thing, failing us all, decade after decade.

One thing that's especially important to remember is that there is no "we" that is causing climate change, even though my language often slips. There are a specific few rich white men who have done the bulk of the damage out of sheer greed. This basic fact is one of the greatest scandals in all of human history. They had no right to do this. The fact that they did should fill you with rage. As the folk singer Utah Phillips said, "The earth is not dying, it is being killed, and those who are killing it have names and addresses." This emergency is personal.

For the past three decades climate change was often considered an unfortunate but unavoidable consequence of civilization—not a by-product of a fundamentally flawed economic system. Now the intersectional nature of our climate emergency is becoming impossible to ignore. Environmental campaigners have focused too much on small changes that maintain the status quo—like recycling programs, redesigned light bulbs, and electric cars—which have only helped further concentrate power in the hands of the people who have benefited from an unjust and punitive system. To date, the conversation focused on adaptation, not transformation—a thinly veiled admission that high-emitting lifestyles weren't up for negotiation: if you don't like the consequences of climate change, the people in power tell us, you should figure out a way to get over it. In reality, climate change is a symptom of a distorted, imperialist worldview: the continued exploitation of the Earth is justifiable because humans exist separate from and can claim dominion over the environment. This worldview, championed by Christians and captains of industries alike, promotes and preserves business as usual and has led us to the brink of ecological collapse. While a majority of people in the US and Europe accept that climate change is happening, most reject the need for radical action. Once this mindset changes, however, once we start to see a different world as not only possible but inevitable, all the millions of incremental changes a carbon-free society requires will follow in a whirlwind.

The next thirty years will feel like a race to catch up in the mismatch between how things seem and how they actually are—a mix of terrifying news broadcasts about the collapse of distant ice sheets, the odd angst of daffodils in January, the grotesque normality of knowing the best places to buy N95 masks to protect your toddler's lungs from wildfire smoke, the panic of not knowing if it's even worth it to save for retirement and not being able to even if you wanted to, the anger as yet another last-chance election tips the wrong way.

In a 2018 essay for On Being, NASA climate scientist Kate Marvel wrote that this moment demands courage, not hope. Unlike hope, which Marvel rightly said is designed to provide comfort during a time that is decidedly uncomfortable, courage demands that we confront the world as it really is: uncertain, chaotic, yet imbued with possibility. "The world we once knew is never coming back," wrote Marvel. "Courage is the resolve to do well without the assurance of a happy ending."

It's increasingly obvious that the old ways of thinking haven't worked. Nothing we are doing is working. Environmentalism, as we know it, has failed. After decades of haggling and setting targets, the world is still belting out carbon dioxide at record rates. To me, that's an indictment of an entire generation. We should no longer just be looking for more ways we can recycle; we need an emergency strategy.

My generation, the millennials, never have known a time when climate change wasn't a grave threat. Back in 1988, when carbon dioxide in the atmosphere first exceeded 350 parts per million, I was still watching *Sesame Street* and digging up worms in the backyard of my family's farm near Topeka. For years environmental activists have told us that we can make progress by tinkering with the status quo, that a big part of halting warming is buying the right car, clothes, and moisturizer; avoiding the dirty products; and reforming the way consumer goods are made. But the world's emissions keep climbing.

If it feels like climate scientists are continually finding that the change we're inflicting on the natural world surpasses our previous worst estimates, you'd be right. In many ways, the planet is changing so fast that we simply don't have time to study the problem anymore. Science isn't meant to be responsive to change this rapid. Our science reports are obsolete even before they're published. Tempered political responses to a planetary problem like this are tantamount to a death sentence on a civilizational scale.

A livable world achieved through slow changes may have been possible in the 1980s, but it's a fantasy now.

It's time to rethink the whole thing.

Ours is a time of unnatural disasters. There is no way to separate our experience of the weather from the decisions we make as individuals and as a collective society. Humanity has become a planetary force, for better or for worse.

In Puerto Rico, following Maria, folks quickly realized that they were on their own. Mutual aid communities came together in the aftermath. Neighbors began to assemble de facto public-works efforts. And slowly, calls rose on the island for a new relationship between the people and their government. Amid all the chaos, communities spontaneously and effectively built a new vision for how to work together to create positive, lasting change. The grassroots organizing that helped spawn Puerto Rico's bottom-up recovery goes back decades. Since Maria, the island has transitioned from the occupation and protection of coastal mangroves, which buffer the strength of storms, to the establishment of locally controlled solar microgrids, which now form the backbone of dozens of communities across the island. A spirit of mutual aid is flourishing in Puerto Rico, which is becoming a model for some places in the mainland United States as well.

"There has been more openness in identifying the relationship as colonial and identifying it as extractive," Marisol LeBrón, an assistant professor in the Department of Mexican American and Latina/o Studies at the University of Texas at Austin, told me. Her work focuses on the Puerto Rico diaspora. "And more critiques towards colonialism, more calls for decolonization."

Instead of thinking of these brave communities as outliers—abandoned by their governments and the prevailing economic system, and surviving against all odds—we should think of them as seeds of a beautifully flourishing new phase of humanity. Diminishing their potential erases what they could become five or ten or thirty years from now. Transformational change can happen fast.

The first step to creating the change we need is imagining that it's possible.

What's the best way to produce large-scale structural and societal change in the broader world, outside the confines of your local community? Is it making radical changes in your own life and leading by example? Is it building a broad group of like-minded and motivated individuals to transform communities and cities? Is it supporting those working tirelessly to radically overhaul the entire system?

According to the IPCC, it's all of this, all at once.

"The decisions we make now about whether we let 1.5 or 2 degrees or more happen will change the world enormously," Heleen de Coninck, a Dutch climate scientist and one of the IPCC special report's lead authors, told the BBC. "The lives of people will never be the same again either way, but we can influence which future we end up with."

If the world's governments took their findings seriously, it would be nothing less than revolutionary—a radical restructuring of human society on our planet.

"We're talking about the kind of crisis that forces us to rethink everything we've known so far on how to build a secure future," Greenpeace International's Kaisa Kosonen told Agence France-Presse in response to a draft of the report. "We have to try to make the impossible possible."

Nothing about this world is permanent. We can demand something better. Our imaginations are the most powerful tools we have to fight climate change. What exactly that future looks like, of course, can't be known for sure. But we do know that it will come to pass—and we will be the ones to create it.

With a problem this daunting and overwhelming, it helps to start small and personalize what the next few decades will look and feel like. For the scientists who've been intimately connected with the data for most of their adult lives, it's probably easy to visualize what the next thirty years will bring: difficult questions from kindergartners as we try to explain why coral reefs are vanishing, tears of joy watching intersecting lines on a graph as the world shuns oil and embraces electricity, the easier-than-expected step to put your chances of tenure on the line and join a protest.

Our lives and actions are not meaningless—in fact, the opposite is true. Our rapidly changing climate proves that we are all connected, that every action we take can alleviate suffering not only tangibly, locally, and immediately but across the vast network of life and throughout all time. Our lives and actions may be temporary and imperfect, but by accepting our tiny place in the world, we unlock a deeper way of being. And we help to bring about a future that will be unimaginably better for everyone.

Here's what must happen in the next three decades for deep cuts in emissions to be possible:

**We must articulate a shared, hopeful vision of the future.**

**We must tear down the current system.**

**We must begin building a new world that works for everyone.**

This timescale and to-do list aren't (just) the outlines of an aspirational progressive utopia; they're scientific imperatives, backed by the IPCC, to preserve human civilization. In order to prevent additional planetary warming, we must reduce emissions by about half of current levels before 2030. This requires a total overhaul of nearly every aspect of human society. In the words of the world's best scientists, the next decade will be defined by transformative change.

For systemic transformation to happen, we need to ask ourselves:

**What is my vision of a better world?**

**What steps need to happen for it to become a reality?**

**Starting today, where do I fit in that vision?**

The good news is the much-needed phase of urgent, creative imagination has already begun.

By 2050, the world will have gone through a period of rapid decarbonization and rebirth—a massive, societal shift designed to preserve life as we know it. The urgency of our situation means that world-changing visions of a better future—the specific people and specific ideas that will shape the next phase of humanity—are already in place; they're just not being heard. If the climate change struggle has failed up to this point, it's only because we've been listening to the wrong people.

If you've been left out of the conversation for the past three decades by the (almost exclusively straight, privileged, and white) men who lead the companies and governments responsible for the climate emergency, you might be so mad at the system you feel paralyzed. That's okay. There's still time for your voice to make a difference. If you've been part of the conversation this whole time, you might be scared of radical change and feel grief or misplaced guilt for even a small role in letting it get this far.

That's okay too. It's time to pass the mic. Systemic change calls for courage on all scales in all places. No matter if you feel ready or not, the world needs you.

Right now the dominant narrative of our climate future focuses on the inevitability of apocalypse. Instead, we have to start talking about the inevitability of a better world. By explicitly focusing on aspirational futures, we make them more likely to actually happen. What will emerge from this time of radical change is unpredictable, but I do know how it will come about: collaboratively, as those who have been systematically excluded from determining their own futures begin to claim power and have their voices heard. These necessary, life-giving visions of our new world will become reality once we tear down the barriers that have been constraining them all along. We can use storytelling as a strategy—a political tool to change the course of history. That is exactly what must happen to emerge from this liminal space.

"From my point of view, it's not just climate change that is the problem," Samantha Earle told me. "You still are then left with a raft of other environmental problems. You're still left with massive inequalities, mass discrimination all over the place. It's not just one or two things that have gone wrong. Our imaginary is systemically problematic. And I think that these things are all related, that it's not just a coincidence that all of these things are rotten."

Earle thinks that the root problem is the concept of ownership.

With ownership comes control, and with control comes hierarchies of power: those who own more can marginalize and take advantage of those who own less. The concept of ownership is connected to all our social struggles: misogyny, racism, colonialism, and the corruption of modern democracy. And, of course, the concept of ownership has directly resulted in the climate emergency, which is really a crisis of overconsumption by those who already control most of the world's resources—with dire consequences for those who own almost nothing. Alternatives to a world built on ownership may feel unfathomable right now, but it's what we'll need to strive for if we're going to survive this century.

It's easy to see climate change as a disaster. But I think the biggest disaster is how we treat one another. Only in an unjust and unequal world can a civilizational-scale threat like climate change fester as it has without so much as a careful thought by those in power. The task of our time is to re-create our world with new power structures that make systemic problems like climate change much less likely.

In a world without the centrality of ownership, we'd need to develop deeper relationships with one another and with the natural world, emphasizing values and actions like consent, care, dialogue, and trust.

We don't have to create that world from scratch. We already know a lot about what that world could look like: it's an idealized version of how we already behave toward our closest friends and family members. We don't own our friends and loved ones; we are in a mutually beneficial relationship with them. Scaled up, that could go a long way toward building a richer, deeper, more meaningful, lasting, and flourishing global society.

Indigenous peoples in the Americas provide centuries-old examples of societies that successfully operate without the modern concept of ownership, surviving some of the worst atrocities in human history. "Look at what colonialism and capitalism and industrialization under partnership together did," Kyle Whyte, a professor of environmental philosophy and ethics at Michigan State, told me. "They undermined not only Indigenous people but everybody else's trusting of each other, their consent with each other, their accountability and their reciprocity with each other."

Whyte is an enrolled member of the Citizen Potawatomi Nation. His work focuses on gathering examples of Indigenous experience with and advocacy for cooperative relationships as a direct challenge to the Western ownership-based status quo. Rather than relying on top-down systems of rules and laws, which tend to be co-opted by people hungry for power, control, and ownership, Whyte champions successful models for self-governing and self-determination that depend on webs of responsibility. When it comes to climate, Whyte said, the key is to nourish those webs of responsibility through dialogue and "kinship" as the underlying framework of a society that can withstand and even encourage radical change. "I don't think that we're ever going to achieve sustainability if we don't have a society where all of the laws, all of the policies are not deeply consensual," he told me. "An environmental problem is one and the same as a problem of consent." Whyte's understanding of consent and kinship is simple and is rooted in international law: asking permission, frequently checking in with one another, patiently building authentic relationships, truly valuing one another as we would with our own family.

Whyte knows the realities of the past and the future are in many ways identical. He knows that his people have seen all this before, and he knows that if we aren't able to reimagine our relationships with one another, unbound catastrophe will happen again.

"Indigenous people are already in the dystopia, at least according to our ancestors, because we have seen the undermining of all these relationships and we've already been through multiple forms of human-caused climate change just in the last two hundred years that are actually far worse than the dire twelve-year predictions. Our response is that we're going to try to work on strengthening those critical relationships, trust, consent, accountability, reciprocity, among others," he told me. "The way the Earth system works, we might not have enough time."

But the worst thing we can do right now, according to Whyte, is panic. That's counter to the messages that many environmentalists and climate campaigners are sending right now. The best climate response is what Whyte is doing: meeting person-to-person, talking about our differences and our shared dreams, and plotting a future together.

"Improving people's behavior and improving the climate involves having all sorts of different types of relationship-building meetings and facilitative discussions and dialogue—that's actually how you solve that problem," he told me. "The legalistic aspects of it, the policy changes and new laws, that's really just the most surface level of change."

Read together, Earle's philosophy and Whyte's kinship advocacy lead to the same conclusion: our world, as it currently is, can no longer last. And there are promising signs—in youth protests, in the rising voices of people of color—that a new world is beginning to emerge.

"When we finally have people who are so different from this Western model challenging it, then finally there is some hope of actually having some traction in a meaningful sense," Earle told me. "If we want something different, it has to be really different."

Whyte's idea of a society built on webs of responsibility—mimicking the ecology of the natural world—seems as good a framework as any I've heard for a basis of what comes next. The challenge will be remembering that, in each and every moment, we have a choice to remain in the old, dying imaginary or leap forward into a new one.

"As for climate change," Earle told me, "if we fail to change, it's going to literally be the end of all imaginaries and all civilization. We have to simultaneously know that there is a man behind the curtain, and that man behind the curtain is all of us, by the way. All of us can change the imaginary, all of us. In fact, we do it all the time, or at least we don't change it, but we reproduce it. But every moment we are faced with the opportunity of either reproducing it or challenging it and rebuilding. I love that aspect of it, because it's radically empowering."

We have to abandon the idea of ownership and instead seek consent with one another and the world we call home. Climate change isn't the problem—it's a symptom of the problem.

* * *

For the first time in the history of humankind we live on a planet where all our daily actions have immediate and permanent physical ramifications for every other living thing we share this world with. For the first time we live on the doorstep of a truly intersectional, interconnected world. For the first time we have to build a world that works for everyone.

Our task this decade, and in all the decades that follow, is to make a future that is radically inclusive, equitable, and just.

The near-simultaneous emergence of the harrowing IPCC special report and a vibrant youth movement calling for climate justice at last makes it possible for us to imagine what a near-future global society could look like that truly works for everyone.

While researching for her book *Why Civil Resistance Works*, Harvard political scientist Erica Chenoweth discovered a surprising and empowering truth about the science of revolution: throughout the twentieth century, every single nonviolent movement to create political change that received active participation from at least 3.5 percent of the population succeeded. Every single one. And many succeeded with many fewer people.

What's more, Chenoweth and her colleagues found that nonviolent movements tend to help foster democracy. That's a clear sign that what the climate movement is doing is already working—they just need your help.

Of course, we will still need people working on technical solutions. Like Greta, I've vowed to stop flying, and I follow a mostly vegan diet. I'm also in the process of converting my entire front and back yards to garden space. But these are just things I'm doing to help motivate me toward bigger changes.

Demanding systemic change from those in power is by far the most important thing any of us can do to help the climate movement, and what that looks like in practice is having conversations with anyone who will listen about how important this moment is. The technical solutions are important, sure, but they will be encouraged and fed faster by an uprising that demands more.

In 2017, Katharine Wilkinson led the ambitious Project Drawdown, which systematically quantified the most effective ways to stop and reverse greenhouse gas emissions globally. The book she wrote about it, *Drawdown*, became a *New York Times* bestseller. Finally it answered the question: What are the most important things we can do to fix the most important problem in history? Tops on the list: valuing, empowering, and caring for women and girls. That may be surprising to some, but not to Wilkinson. For too long, she said, we've taken a technocratic approach to climate change, expecting the right mix of taxes and regulations to move the invisible hand of the market to prioritize solar power over coal. The time for that approach is over.

Wilkinson told me the biggest success of *Drawdown* was "the very ability to begin bringing more emotional content into a conversation that is fundamentally about what it means to be human, and who we are and what we're doing and why any of it matters. I don't know any way to come to those conversations with just science and some tech solves. To me that feels like a real sort of escapism from the questions that the Earth is asking us to wrangle with. This idea that the best means and frameworks for problem-solving are coming out of business. Which is just freaking nonsensical. There are so many truths that are not even making it to the table for consideration as we think about what to do and how to do it."

That leaves the emergence of a resilient, relentless global movement to demand that these plans turn into reality.

Going forward, the climate movement will continue to morph into a social movement and ally itself with other social movements—that's the only way the necessary deep changes will persist long enough to be effective. The values that will bring us to a sustainable future are consistent with the shared values across cultures and the core of what makes us human.

These are examples the climate movement can follow to imagine a better future. By recognizing race and gender as social constructs, we can start to imagine aspirational futures—prophecies of a better world. When we forget that, we end up with things like cities that aren't built for people with disabilities, a justice system that's discriminatory, and a global economy that's systematically destroying the ecosystems that make it possible.

It's true that we've come much further toward climate catastrophe than we should have, but it's also true that change is finally starting to happen on the scale that's necessary to change everything.

In the US, the country with the greatest historical responsibility for climate change, youth took Greta's lead and went to the streets starting in 2018 to demand a Green New Deal, with the idea that only a total reimagining of the social contract—including jobs, health care, and housing for all—would allow citizens to unleash their creativity and work together to build the world we need. Voters are inspired by their leadership—some polls show that climate change has become the number one issue in the run-up to the 2020 presidential campaign. At last the US is ready to talk about radical change.

Because women and people of color are leading the development of the Green New Deal, informed by surviving centuries of existential threats, there's a much better chance that such a policy change will result in a better future for everyone. On her first day in Washington, DC, as a newly elected member of Congress, during a youth-led sit-in in the office of Speaker of the House Nancy Pelosi, Alexandria Ocasio-Cortez said that only a bold reaffirmation of the basic social contract in the US could avert climate catastrophe. A few months later, she expanded on this vision: "Our extractive, wasteful, fossil fuel economy is posing a direct threat to our own lives. There is a better way: one that's conscious, just and prosperous. We will not be able to save our planet without first changing ourselves. That is the task before us."

Her words became fuel for a youth-powered movement to help advance a Green New Deal—one of the boldest visions of the future in US history.

As the Green New Deal evolves, one thing is for sure: its vision is the best chance we've ever had to lay the groundwork for an ecological society that puts justice, consent, and equity at the center of a flourishing new stage of human civilization. Drawing from long-standing demands of the environmental justice movement and the Movement for Black Lives, and from

American values dating back as far as the Constitution, the Green New Deal has become a clarion call for real change.

Similarly, the Sunrise Movement, the youth-led organization at the forefront of championing a Green New Deal, is boldly optimistic that its vision is achievable. If they're right, we're at the dawn of a new era. We're going to have a cultural wake-up call, and things are going to be flipped on their heads. So far, the group has delivered. In less than a year, the Sunrise Movement—led by a diverse group of young women and men and championed by Ocasio-Cortez—has been able to influence the entire national conversation, and has almost single-handedly pushed climate change to the heart of US politics.

Varshini Prakash, cofounder and executive director of the Sunrise Movement, is almost an embodiment of the movement's values. She's smart, idealistic, and expects to win. Her enthusiasm is infectious.

"We're not just putting out a Green New Deal to be shocking," she told me. "We're putting out the Green New Deal because literally thousands upon thousands of extremely smart climate scientists told us that unless we make unprecedented changes to every part of our economy and society over the next twelve years, we're screwed."

She also is blunt, swears a lot, and—increasingly—wields a lot of political power. Prakash may not have set out to be this successful at her job, but in the midst of a political moment when almost anything could happen, she and the Sunrise Movement are well positioned to have great influence on the direction the US—and the rest of the world—takes over the next decade.

"What we're actually doing is, we're preparing. We're preparing for what's happening rather than acting like it's not gonna come and acting like it doesn't exist. That's what the Green New Deal is all about."

Prakash is a visionary at a time when visionaries are needed in multitudes. The main difference is, she's making it happen, and people are listening.

"There's this whole move towards an actually caring society and an economy that actually works for all people, that doesn't leave people behind, that doesn't treat some as disposable and others as über-important. What is so audacious or terrible about just calling for something that allows for everybody on this planet to thrive and exist?"

On a pleasant spring evening on the campus of Macalester College in Saint Paul, I sat in on a Sunrise Movement town hall meeting, just as the 2020 presidential campaign was beginning to heat up. Less than a mile away, the Mississippi River was cresting at one of the highest levels ever recorded in Minnesota, part of an unprecedented spring flood that spanned more than a dozen states across the Midwest.

Most of the people in attendance were older and from the neighborhood. For the first hour, all the presenters were students. This was, it felt, an intergenerational transfer of knowledge and experience.

One of the first speakers in the auditorium was Dakota McKnight, a freshman who spoke of his anxiety and persistent feeling of powerlessness in the face of news that "civilization could crumble in our lifetime."

Looking around the room at this, I saw several heads of all ages were nodding.

"It's not too late. Your life is not worthless," he said. After joining in the sit-in at Pelosi's office the previous fall, "for the first time in a long time, I saw hope. That hope has allowed me to squash anxiety."

The Green New Deal itself is a message from the future. On one of the transition slides of their PowerPoint presentation, the organizers of the event made a promise that this generation will be the ones to build a new phase of humanity: "We will destroy out of love, and create out of anger. We can pull off the unimaginable."

One of the later speakers was Christie Manning, a Macalester environmental psychologist, who told the students that her own research shows that in a world where the vast majority of our elected representatives and other elites in society might privately agree with the climate movement but publicly oppose action, social movements like Sunrise are prime to create rapid change. The most effective way to break through this dissonance, she said, is storytelling, because it personalizes the problem and makes it feel more immediate, which physically changes your brain chemistry and is more likely to generate empathy. "Your story can change the debate," said Manning. "It's uncomfortable, but it's what's asked of us. Speak, and speak publicly with your friends and people in power."

Leaving the event, I became more convinced than ever in the power of story, and the idea that sharing stories could be an effective strategy that's simple and radical at the same time.

Prakash told me that part of what makes Sunrise so important is that it prioritizes the calls for greater ambition coming from people of color and those on the front lines of climate change. "There are people calling for a carbon tax that doesn't get us to 40 percent emissions reductions by 2040. That's fine, but think about who you're leaving behind when you call for those kinds of things. Communities of color don't have that kind of luxury to call for that. It's not what's realistic right now because those communities will drown. That's why '1.5 To Stay Alive' was such a rallying cry during a lot of these UN climate conferences. It feels super basic, but I actually think there's something really radical in saying our economy, our government, should serve all people. If that is a basic tenet that we believe is true, then everything about the way that we make our policy decisions, the type of research we conduct, the kinds of economists we trust, everything would change about it. I just think that's really fundamental."

The solutions, too, are led by people of color, and that makes them revolutionary.

"I think something that people sometimes do is talk about these ideas like they're very new. But they've been in existence for a while. Indigenous communities have acted like this and treated the Earth in this way for a long time before the effects of colonization really destroyed so much of that. Cooperative economies based in mutuality were created around the Civil Rights era, largely in response to the lack of resources and support that Black people received post-Reconstruction. A lot of that is based in why we're even calling for the things that we're calling for."

The Green New Deal's call for historic, transformative change—at an emergency pace—could see the US kick-start a new era of responsible climate policy, "a new national, social, industrial, and economic mobilization on a scale not seen since World War II," according to the resolution introduced into Congress.

Its ten-year plan would provide "100 percent of the power demand in the United States through clean, renewable, and zero-emission energy sources" and a "just transition for all communities and workers." This is likely at the limits of technical feasibility, even with the hedge of "net-zero" emissions, which would allow for a slower complete phase-out of fossil fuels.

Simply put, it's a manifesto for a restructuring of society to thrive in the climate change era. The Green New Deal would address "systemic injustices" head-on in frontline and vulnerable communities through a living-wage job guarantee, public education, universal health care, universal housing, and "repairing historic oppression," all the while promoting a resurgence in community-led democratic principles.

"I think that this is a very special moment," Ocasio-Cortez said on the day the Green New Deal was launched. "We have a responsibility to . . . show what another America looks like."

Paying for it would likely require trillions of dollars of spending—which could be achieved as simply as just making sure rich people pay the taxes they already owe—but could be accompanied by a restructuring of the US economy that would sharply reduce inequality in the process. Polling for early versions of the plan showed overwhelming support from the public, even among Republicans. There's a real chance that this is the policy framework that will deliver the future we need. In the context of our ongoing planetary emergency and the United States' long struggle to productively confront climate change, it's impossible not to see this as an investment in the future of our country, an investment in the stability of the planet and the survival of human civilization.

And now it's up to us. We've got to do whatever it takes because this is our last, best chance at the world we need.

## WHERE COULD OUR STORY GO FROM HERE?

At the dawn of a new decade, we find ourselves with a terrifying and fleeting opportunity to reshape our world. A rapidly growing movement of people is taking the latest science very seriously and strategically organizing a revolutionary vision. Chenoweth's research shows that radical change is easier to achieve than most people think: her 3.5 percent rule, proved across decades of history in dozens of countries across the world, says that a nonviolent mass movement will succeed in the United States with the active participation of just 11 million people. The seeds of this revolution have already been planted. In September 2019, on the largest day of climate protest in history up to that point, New Zealand became the first country to reach the 3.5 percent mark. Weeks later, the country's prime minister signed into law one of the most ambitious national climate plans in the world.

A culture of radical change is not a best-case scenario; it is what absolutely must happen for us to preserve a habitable planet. Somehow everything we know is going to have to change, and at a pace most of us can't fathom right now. With those kinds of stakes, giving up is unacceptable. So we might as well work together to imagine an irresistible, justice-filled world that we're excited about bringing into being. That's what I'm trying to do in this book: sketch an outline of what that change might feel like, decade by decade, until we have built a world that can truly flourish.

This book is divided into three chapters: the 2020s, the 2030s, and the 2040s. In each, I briefly step into the future and report on what it will look and feel like to be a part of radical change on the scale the science says is necessary to maintain a stable climate. This speculation is grounded in my analysis of countless scientific studies and dozens of interviews with people on the front lines of change from all over the world. I have tried to anticipate how the physical constraints of our oceans, lands, and atmosphere might work both with us and against us. And most important, I have tried to show how the work we must do is not just about the science; it's about learning how to care for one another again. This book is an example of what it might look like when we win.

In 2020, there's a growing realization that there is no choice left but to demand a better world. Humanity is looking extinction in the eye and deciding to thrive instead.

The strategy is simple. By growing the movement to include people who've been systematically excluded, radical climate action is increasingly inevitable. But that doesn't mean there aren't significant headwinds. There's a dangerous escalation of rhetoric. Anything could happen. Amid this political, economic, and environmental uncertainty is a rush for power and control. A rise in nationalist, populist, white supremacist ideologies has already provoked a dystopic response that some are calling eco-fascism—which has included border closures and an escalation of inequality, resulting in a growing sense of desperation. Centrists and their allies have become de facto agents of the status quo. The pro-growth establishment, aligned with business and industry, is what got us into this mess in the first place, after all.

The only moral and ethically just response to a situation like this is to fashion an entirely new system.

Those fighting for climate justice, of course, have science on their side. The 2018 IPCC special report imagined possible paths for the 2020s that could be straight out of a manifesto for socialist utopia. Via either a combined surge in realization of our mutual duty to one another or alarmingly rapid developments in the severity of climate change itself, the 2020s will be a decade of rapid, radical change.

The IPCC's "best-case" scenario for this century imagines "strong participation in all major world regions at the national, state and/or city levels" beginning in the 2020s, leading to a rapid phase-out of internal combustion vehicles and a massive scale-up in sourcing bioenergy—for carbon capture and storage—from agricultural waste, algae, and kelp. Ecosystem-based economic practices, large-scale reforestation, and changing attitudes all favor plant-based diets and cities where life is intertwined with nature, not separate from it.

The IPCC report also imagines a worst-case scenario: not much happens in the 2020s. If the nations of the world fail to keep their promises, or if our attention to climate change falls out of fashion, we'll have a world that, by the end of the century, "is no longer recognizable." The details the IPCC imagines of that world are harrowing: "Droughts and stress on water resources renders agriculture economically unviable in some regions . . . Major conflicts take place. Almost all ecosystems experience irreversible impacts, species extinction rates are high in all regions, forest fires escalate, and biodiversity strongly decreases . . . These losses exacerbate poverty and reduce quality of life. Life for many indigenous and rural groups becomes untenable in their ancestral lands. The retreat of the West Antarctic ice sheet accelerates, leading to more rapid sea-level rise. Several small island states give up hope of survival in their locations and look to an increasingly fragmented global community for refuge. . . . The general health and well-being of people is substantially reduced compared to the conditions in 2020 and continues to worsen over the following decades."

In every situation, the 2020s hold massive importance for our collective future for centuries to come.

No matter what happens in the next three decades, we must answer fundamental questions of how to build a better world with detailed plans that include a diversity of voices from every nation on Earth. We have only one shot at this.

**How will we remake the world's food systems?**

**How will we retrofit every home, office, and factory in the world?**

**How will we rebuild our cities to make sure they can withstand more extreme weather?**

**How will we reimagine transportation in a world where fossil fuels are being rapidly phased out?**

**How will we reform our democracies to make sure all this is done with justice and equity and respect for everyone, especially historically oppressed peoples?**

**How will we transform our global economy to be regenerative rather than extractive?**

**How will we make enormous headway on all of this as soon as possible—against the most powerful industry in human history and their allies at the highest levels of government?**

Writing this book has been oddly liberating. It has become easier to make forward-thinking decisions: eliminating air travel, commuting by bicycle, raising my kids to be mostly vegan, teaching them to consider the consequences of their actions.

The science on climate change is brutal and unforgiving. Returning the Earth's atmosphere to 350 parts per million of carbon dioxide, perhaps infeasible with current technology, isn't the right goal. The aim of climate activism isn't to be ideologically pure, nor is it to erase the sins of the previous generations; it's to ensure that future generations are handed a world that isn't on the verge of going to hell. The goal isn't activism, it's survival. Without a truly radical remaking of how we treat one another, a better world won't happen in my lifetime. And if it doesn't happen in my lifetime, then it's almost certain that future generations won't get another chance. With science as a basis, my assumption in all my work is that radical change is now inevitable.

Lately I've been writing love letters to the future. Literal letters, addressed to people I love, tapping into my unrestrained idealism. I've been learning that I know how to do this after all. The next few decades are going to feel like falling in love—setting aside everything you thought you knew and trusting that you'll end up in a radically different place you never could have achieved on your own.

We have reached a unique and vulnerable moment in human history: our futures are simultaneously dependent on the actions of others and defined by our personal daily choices. This reality demands that we interact with one another, that we come together—for our own personal survival and the survival of life as we know it on our planet.

What would this world actually look like, the world where we win back our future?

This book is my effort to produce a radical vision for our future Earth. This book is my love letter to the world. And it's my invitation to you to write your own revolutionary love letters. Like all revolutions, I hope this conversation goes on for a long, long time.

# PART II

# 2020–2030:
# CATASTROPHIC SUCCESS

*I took a video on my phone because I thought that was the end of all of us. I took a video and put my phone in a dry bag with my ID and my wallet. I thought, "This is going to float and somebody's going to find it." To give them an idea of what it was like during the storm. I wasn't thinking at all about, if we take this picture here, more people will see this. I wasn't thinking too much about how it was going to be seen, I just thought, let me give people an idea of what's going on here at this time. I was afraid that anything could happen.*

—John "Junior" Rulmal, in 2015, recalling when Typhoon Maysak hit his home island of Ulithi, part of the Federated States of Micronesia

The first human inhabitants of the Marshall Islands were pathfinders, migrating between remote islands with an intimate knowledge of the winds and weather of the Pacific. They were among the best sailors who ever lived, with an incredible connection to the wind and the ocean. Their intimate knowledge of the world around them allowed them to thrive amid impossible odds. Now, more than a millennium later, the weather has become an enemy of their descendants.

Buoyed by thousands of years of rootedness on these coral fringes, in some places only a few feet wide, they refuse to be cast aside, or forgotten to history. Their experience is forcing once-unimaginable questions: What would it take to leave the place you call home? And what does it mean when you lose the place where your ancestors lived—the place that literally defines you? What does it mean to know your home will be annihilated? What does it mean to decide to stay and fight anyway?

Halfway between Hawaii and Australia, Micronesia—the name given to a patch of ocean twice the size of the United States—includes the Marshall Islands, Palau, Kiribati, Nauru, the Federated States of Micronesia, and three US territories: the Northern Mariana Islands, Guam, and Wake Island. These nations and territories total just one thousand square miles of land, less than half the size of Delaware. (A quarter of that total land area is the main island of Guam, Micronesia's largest land mass, itself roughly the size of Chicago.) About half a million people live on the two thousand islands of Micronesia, a constellation of safe harbors across a huge watery vastness.

The Marshall Islands is not a small island state, it's a large ocean state. Just offshore, the reefs of the Marshall Islands contain some of the most productive and biodiverse waters in the world. The islands' 29 coral atolls contain 1,156 individual islands spread across a stretch of the Pacific as wide as the distance from Texas to North Dakota, an oceanic supercontinent. The islands have a maximum elevation of just thirty-two feet above sea level, though the vast majority of the land here lies less than six feet above the tides—exactly the amount that global oceans are expected to rise this century.

This isn't the first time the Marshall Islands have faced annihilation. Just one lifetime ago, during the Cold War, these islands were used as target practice for US nuclear weapons. A containment dome still sits on Runit Island, designed to sequester radioactive material from nearby fishing grounds for centuries. It wasn't designed for rising sea levels, though.

And so the Marshall Islands have long been the epicenter of a global resistance movement, well before their successful effort to convince the world to aspire to a warming target of 1.5 degrees Celsius at the Paris climate talks in 2015.

With the signs of rising seas and increasingly extreme weather, some Pacific natives have started to think of themselves and their homelands differently, holding out hope that truly radical action would be enough to turn back time and reverse some of the damage that's already been done.

In 2013, when Typhoon Haiyan roared through the tropical Pacific, it brought with it a new era. Before hitting the Philippines, Haiyan moved through Micronesia, passing as close as five miles offshore from the tiny island of Kayangel in Palau. The strongest winds in a tropical cyclone circulate around the central eye, a cloud-free region of descending air just a few miles wide caused by outflow from a ring of very intense thunderstorms. Haiyan's traverse near Kayangel brought winds strong enough to devastate the island. Reporting for the Solutions Journalism Network, Ari Daniel said that four families survived the storm by crowding into the only place on the island with concrete walls: the hospital bathroom. After the storm passed, local officials evacuated all 138 people who lived on the island and abandoned it for months while repairs could be made. Haiyan was a superstorm—the strongest tropical cyclone to make landfall in recorded world history. It hit with estimated 195-mile-per-hour sustained winds—and I say estimated because no weather station could survive such ferocity. That figure is derived from weather satellite estimates of the temperature of Haiyan's cloud tops—a proxy for the vigorousness of its thunderstorm activity (taller thunderstorms have stronger winds and colder signatures as seen from space). Haiyan maxed out the most commonly used satellite-based intensity measurement, exceeding the theoretical maximum strength of a tropical cyclone as conceived by the meteorologists who invented the scale. Using the five-tier US classification system for hurricanes, which grades storms based on their maximum wind speed—the easiest part of a tropical cyclone to measure because rainfall and storm surge can vary strongly with local topography—Haiyan would have been a Category 6.

By all accounts, Haiyan utterly transformed the region. In the immediate aftermath, BBC described "a wasteland of mud and debris." The Philippines' lead climate change negotiator, Yeb Saño, was in Warsaw, Poland, at the time, attending the annual summit of world leaders working toward a global agreement to limit human interference with the climate system. In a tearful address to the other delegates, Saño was defiant. "We may have ratified our own doom," he said. "We refuse, as a nation, to accept a future where super typhoons like Haiyan become a way of life. We refuse to accept that running away from storms, evacuating our families, suffering the devastation and misery, counting our dead, become a way of life. We simply refuse to."

To use a phrase from the Italian scholar Antonio Gramsci, those on the front lines of climate change have "pessimism of the intellect, but optimism of the will." Speaking out against the slow pace of the international effort to combat climate change, Saño began a voluntary fast and unwittingly started a social movement. By the end of the thirteen-day meeting, hundreds of thousands of people pledged their support from around the world. The combination of the unseen ferocity of Haiyan and Saño's selfless words and actions forced the urgency of the changing weather to the forefront of the conversation about climate change and our collective consciousness. He also laid the groundwork for the first-ever global climate agreement in Paris two years later.

On its own, the Paris climate agreement won't be enough to prevent the climate-related demise of Saño's Philippines or the Marshall Islands. With the stakes so high, and sea levels continuing to rise, the dominant narrative of these places, as told by outsiders, is that they are the first nation-casualties of climate change. Within this narrative, the Marshall Islands and the Philippines aren't considered actual places; they are a metaphor for humanity itself, a warning straight out of the Old Testament. Once pure and unspoiled, they have been unjustly sentenced to oblivion, a harbinger of something worse still to come. Within this narrative, no happy ending is possible; failure is inevitable, a fate sealed to history, lost to the rising tides.

That's not how twenty-two-year-old Selina Leem tells the narrative of her birthplace. In the Marshall Islands, she witnessed what was happening with her own eyes. Standing on the narrowest part of Majuro Atoll, her home island, she felt the water crowding in around her with renewed urgency.

During a strong storm one afternoon, she happened to look out her window and saw the waves crashing on her grandparents' graves—the first time she felt personally insulted by the ocean. At that moment, she knew she could no longer be silent.

"I kept thinking the whole world was turning against us, and we haven't contributed anything and we're the ones suffering," Selina told me. "I was really angry at how blind and ignorant the rest of the world was. We basically have to somehow go above that to make them do something." Faced with an impossible task, Selina quickly rose to global prominence as a leading moral voice on a rapidly changing planet.

When those of us in rich countries think about Micronesia at all, we probably think of the warming seas and increasingly extreme weather as something that is happening to people who have no voice or power to direct their own fate. Selina knows better. In that sense, she's already learned what everyone on Earth will need to understand quickly: that we have entered an era when we all are connected by our changing atmosphere, and that a duality of life and death has already defined our shared cultural moment. By our daily actions, we are changing the living conditions for the next generations—for better or for worse. The implications of that knowledge could inspire a whole new understanding of what it means to be alive on this planet.

As an eighteen-year-old Marshallese delegate to the 2015 Paris climate summit, Selina watched as the world debated phrases and sentences that would endanger or preserve her homeland's existence within her lifetime. In the waning hours of the conference, tensions rose between the United States and China—the two leading emitters—and a loose coalition of dozens of the world's most vulnerable countries sprang up to try to stage a last-ditch effort to prevent the talks from collapsing. Tony deBrum, who was then the Marshall Islands' foreign minister and the head of their negotiating delegation, assembled a "high ambition coalition." Led by DeBrum, the coalition ultimately won a stronger agreement than anyone had been expecting. The Paris summit culminated in the world's first agreement to reduce emissions of the pollutants that are making weather more extreme and causing the oceans to rise.

In Paris, DeBrum asked Selina to tell her story—and make a case for her country's very existence. In her speech, Selina didn't shrink from this reality. After introducing herself to dozens of heads of state and delegates from 196 nations as a "small island girl with big dreams," she recalled her home's vulnerability, standing on the atoll she grew up on: "On my left is water, on my right is water. I am surrounded by water." She said she only began to be afraid of the water after her grandfather told her about the ice melting at the poles when she was six or seven, which at the time seemed to her like a horror story.

"Sometimes when you want to make a change, then it is necessary to turn the world upside down," Selina said. "This agreement should be the turning point in our story, a turning point for all of us."

As Selina spoke, she held up a strand of coconut husk. "The coconut leaf I wear on my hair and I hold up in my hand is from the Marshall Islands. . . . I hope you keep it and show it to your children and your grandchildren, and tell them a new story, about how you helped a little island and the whole world today."

Though she was just a teenager at the time, Selina spoke with the fiery wisdom of someone who had already seen too much. While the seas grew angrier and more insistent, so too did Selina—and millions of other people from island homes around the world who refused to watch the oceans swallow their homelands.

"If we do have to lose our islands, then we are right now no longer just fighting for the Marshalls; we're fighting for the rest of the world. As I've come to realize, the Marshall Islands are not the only vulnerable country affected by climate change. There's so many other communities, so many other countries, so many other societies around the world that are also affected," Selina told me. "So when we're fighting for this, we're not just fighting for ourselves; we're also fighting for those other people."

After she returned home from Paris, Selina received a flood of responses from people all over the world, some intensely negative, some dismissive of her concerns—even as they claimed to be supporters. One person said that although they agree the world should take action on climate change to help save her islands, "we just can't afford it." During another event in Canada a few months later, the woman introducing Selina solemnly said that many of the flags on the stage might not be seen in the next few years. The crowd silently nodded.

Selina was shocked. "It just hit me. I was like, wow, the rest of the world is already saying goodbye. I just sat back and thought, *What is all this advocating for? What is the role of us Marshallese and us islanders, going around and telling the world that we still want our islands to be there?* Yet it's already very obvious from the woman's response and from the crowd, the way they all accepted it very solemnly, that it's already going to happen. No matter what we do, it's still going to disappear. There are moments like this where I really just want to start yelling and pointing fingers. How many people have already stood on this stage where I am standing? How many more people are going to be crying here, and their pleas gone to ears where no one listens? We are not ready to say goodbye."

Despite the inevitability of catastrophe, they refuse to be annihilated.

Still, storms increasingly batter the islands' coastlines and break down the coral fringes and seawalls that protect the graves of her ancestors. Within those waves, Selina feels a visceral connection to the rest of the world. She recognizes the faces of the first coal barons and oil wildcatters. She sees the busy modern highways and smoking factories in far-off cities she'll probably never visit. She sees your face and mine—people who've grown up amid the comforts of a fossil-fuel-powered economy yet who've never heard of her islands or the fear and nervousness that the rising seas are bringing to her friends and family. With the seeming force of an entire planet, each high tide brings more of the ocean ashore—a watery message of contempt.

In Paris, Selina gave voice to what everyone was thinking: *If the world embarks on a path that essentially sacrifices the existence of entire nations as a negotiating point, where will future leaders draw the line?* Her voice was clear: "This agreement is for those of us whose identity, whose culture, whose ancestors, whose whole being, is bound to their lands. . . . If this is a story about our islands, it is a story for the whole world."

For some in the room, it was the most memorable moment of that historic meeting. Selina's speech was met with a standing ovation.

In the weeks before the Paris climate summit, a graphic photo of a young Syrian boy—drowned on a beach in Turkey after a harrowing attempt by his family to escape a nightmarish war—overwhelmed the world's senses like a punch to the gut. The moment was a deeply personal, tragic, and urgent slap in the face for those who had been paying little attention to what has become Earth's largest forced mass movement of people since World War II.

Multiple studies have now shown that the Syrian crisis was triggered in part because of the fallout and mismanagement following one of the worst droughts in centuries—linked to shifting rainfall patterns due to a warming planet. Along with the pure and immediate horror that humanized a refugee crisis many people knew of only via statistics, the photo said clearly that it was not just rising seas but also the loss of agricultural areas that might redefine a world where rapidly changing weather conditions are beginning to have profound consequences.

The Pentagon has warned that shifting patterns of droughts, heat waves, and melting ice have already become one of the planet's biggest security threats. In some parts of the world, like Syria, this has helped to spark brutal wars and forced people to permanently abandon the only places they've ever known. The Syrian conflict has driven much of the current round of mass migration, but it also foreshadows something far worse: by the middle of the twenty-first century, the UN estimates that more than 250 million people worldwide will be forced to move away from environmentally vulnerable parts of the world if nothing changes.

Shortly after the photo of the Syrian boy emerged, Tony deBrum set the stakes high on behalf of the world's front-line climate nations: "Displacement of populations and destruction of cultural language and tradition is equivalent in our minds to genocide." In the months following Paris, a delegate from Tonga summed up the mindset in the islands from here on out: "In 10 years we drown. . . . Until then, we work."

This century will unfold astride intertwined planetary and human tipping points. Because the weather is now political, it has generated a social movement. Instead of getting lost in the horror of existential change seemingly beyond their control, people like Selina have helped transform the Marshall Islands, along with other countries on the front lines of the climate emergency, into a place of courage. Selina's speech in Paris also signaled the beginning of a global shift of power: the century's middle decades will be guided by the moral authority of youth and those living at the front lines of climate change—demanding that their voices be heard. For Selina's generation, what's happening is much more than changes in the weather and increasingly persistent tides. It's about working together to create a new world that is more peaceful, prosperous, and fair.

* * *

Incontrovertible evidence of human influence on rainfall, temperature, sea level, and cloud cover means that our daily experience of being alive on our planet is now different than it has been throughout the entirety of the hundreds of thousands of years of modern human existence, and is being actively mediated by people in power. That daily reality—and the continued ability of humanity to thrive on this planet—is now subject to the choices we make as individuals, as communities, and as a broader society. Since weather affects almost everything we do, everything from food production to transportation to public health to the very viability of our cities, the fact that human activities are fundamentally changing how the weather operates will create winners and losers. Weather is, more than ever before, a matter of social justice. The atmosphere is now both a weapon and a source of life, and the way we talk about our new shared reality will either empower the communities who stand to lose everything or risk further tilting the scale in favor of the people who will profit from continued business as usual.

As with all political topics, words matter, and the way we talk about the weather matters now more than ever.

International politics has long thought of climate migrants as a problem to be managed, not as a civil rights crisis, stripping families of their humanity during a vulnerable time. There still is, officially, no such legal thing as a "climate change refugee"—the United Nations does not recognize the atmosphere or the environment as an entity that can inflict harm that would qualify a person for refugee status.

During a presentation to the UN Security Council on climate and migration, Michael Gerrard, founder and director of the Sabin Center for Climate Change Law at Columbia University, outlined options for action available under Article 39 in Chapter VII of the UN Charter. The article states that the council "shall determine the existence of any threat to the peace, breach of the peace, or act of aggression and shall make recommendations, or decide what measures shall be taken . . . to maintain or restore international peace and security."

The council, Gerrard said, is able to evaluate whether climate-related displacement poses a genuine threat to peace and, if so, initiate plans for minimizing and coping with large-scale refugee issues. When the council first debated the broader risks of climate change in 2007, including forced migration, it was a controversial decision. Developing countries feared the council might not adequately represent their concerns. Now, though, it has become increasingly accepted that small island states, in particular, face an existential risk due to climate change and that the Security Council could play a key role in encouraging action. "This effort could also spark recognition of the need for significantly greater efforts at mitigation," Gerrard told me. "Climate change offers the Security Council the opportunity to be proactive in preventing threats to peace."

At the same time, the specter of a world pushed to the brink by a surge of climate migrants demanded an anticipation of the special rights of the displaced, the obligations of high-emitting countries to facilitate resettlement, and enforcement of these rights and obligations by the international community. Even though greenhouse gas emissions unequivocally cause harm, it is impossible to assign blame to an individual act resulting in a specific forced migration.

That makes for "a wicked problem," Jessica Wentz, Gerrard's colleague at the Sabin Center, told me. She said it at least partially explains why the international community has been reluctant to take up this issue in any meaningful way. Wentz believes a new protected status applying specifically to environmental migrants might help secure the rights of people forcibly displaced in the future by rising seas or megadroughts. Such a protected status could eventually provide a pathway to citizenship in a foreign country following a climate-related disaster or a slow-onset event, such as sea-level rise.

Along with the leaders of other poor and vulnerable countries, the president of central Pacific island nation Kiribati, Anote Tong, called for a global system of reparations that would take into account the loss and damage climate change is already causing. At the Paris summit, representatives of Kiribati and Fiji announced an agreement that, in principle, allowed the more than one hundred thousand residents of Kiribati to attain residency in Fiji in the event rising sea levels make their homes uninhabitable. In Bangladesh, where an estimated two hundred thousand people are made homeless by erosion each year, the country started a bold plan to reclaim land from the surrounding waters to aid in resettlement. Meanwhile, future migrants fleeing rising seas in the Maldives and Tuvalu may not have a homeland to return to.

From a legal perspective, the looming crisis raises an interesting, though morbid, question: What happens when these island nations, for all practical purposes, cease to be? Under current international law, a country's exclusive economic zone—for which it retains rights to benefit from fishing, mineral exploration, and tourism, for example—is measured from its coastline. If an island disappears because of sea-level rise, will its economy also be wiped from the map? What happens when saltwater intrusion into porous coral soils makes an island effectively uninhabitable? If a place has the appearance of impermanence, how long before the world will see it as already gone? After all, the loss of a place doesn't necessarily happen the day the island goes underwater—an economic exodus may begin decades in advance. Will citizens of a former island nation, scattered throughout the world, still be able to advocate as a collective within the United Nations?

These are the sorts of questions that keep Gerrard and his colleagues up at night. "I think the countries of the world need to start thinking seriously about how many people they're going to take in," Gerrard told me. "The current horrific situation in

Europe is a fraction of what's going to be caused by climate change."

Gerrard has devised an interesting proposal: this century's climate migrants should be provided permanent residency abroad in a manner proportionate to historical national emissions. That means the United States, which holds the dubious honor of being the world's largest historical emitter of greenhouse gases, would be on the hook for millions of displaced people. But considering the anti-immigration rhetoric that has emerged during recent years, voluntary policies that help provide safe harbor for many orders of magnitude more from inundated Pacific islands or parched Sudanese farmland seem almost unthinkable.

In an op-ed published by *The Washington Post* in 2015, Gerrard made a forceful case that the United States bears a unique moral responsibility to confront the climate migrant crisis with a compassionate and welcoming resettlement policy. "International law recognizes that if pollution crosses national borders, the country where it originated is responsible for the damages," he wrote. "That affirms what we all learned in the schoolyard: If you make a mess, you clean it up." Under a worst-case-scenario estimate, the US would be responsible for housing a whopping 67 million people during the next thirty years, more than 20 percent of its current population. The best way to preclude such potentialities, Gerrard told me, "is radical, rapid reductions in greenhouse gas emissions."

The Marshall Islands has been here before: when some islands were forcibly depopulated and used as nuclear test sites during the early years of the Cold War—sixty-seven bombs were detonated in total, the land rendered uninhabitable for generations. Those wounds are still fresh in the minds of many Marshallese.

About one-third of the Marshall Islands' population of seventy thousand now live abroad. Many chose the United States as a consequence of a compact of free association established in the aftermath of the testing that allows Marshallese citizens to move to the US unimpeded.

If you look closely at threatened places like the Marshall Islands, there is a simultaneous apocalypse and cultural rebirth. With so many people already living apart from their homeland, there remains a strong push to preserve traditional cultural practices, like medicines, weaving, and celebrations—a way of knowing the world that is intimately tied to the plants and animals and weather of the islands.

According to the experiences of people I've spoken with in the Marshall Islands, almost no one leaving will list weather or climate as the main reason for their move. And many of them didn't "flee" or consider themselves "refugees"—they are regular people who wanted to improve their lives.

Kianna Juda-Angelo has tried to imagine and build a Marshall Islands that will not only survive but thrive. She was born there but raised in Oregon—she was adopted as an infant and only recently reconnected with her Marshallese family. That identity, found later in life, transformed her outlook and buoyed her hope.

"I get a lot of people that say, 'Why should I give to your nonprofit if you're going to be doing work in a place that's going underwater?' And I say, specifically to other people here on the West Coast: 'Any day we are due for a really big earthquake, and that earthquake is going to put homes underwater and flatten homes with landslides. But we're still here, and we still build on our fault lines, and we will build in the craziest places in Oregon because it's one of the most beautiful places—we love the outdoors, we love the trees, we love the mountains, the list goes on and on and on.' I remind people about their own surroundings first, and then we can get into the conversation easier: there are a lot of people that don't want to move from the Marshalls."

Kianna told me it's impossible to think of our world as a place that doesn't include the Marshall Islands. And that's forced her to be forward-thinking. "I'm actually moving back. My family and I will be moving back, and we'll be building a place there.

"We have to switch our way of thinking to: There's going to be people who want to live there. There's going to be people who come back, like me, so how is it that we can work with the environment? I'm not going to force all the Oregonians to move. We're not going to force the people from the Bay Area to get out because the Big One is going to be devastating. How many earthquakes have we gone through, and we still haven't moved? How many floods has the South in the United States gone through, and they're still not moving, right? Nobody's moving! It's so easy for outsiders to categorize a whole group of people and label them as dumb. Why would you want to rebuild New Orleans, a sinking city? New York City is also an island. The subway system there is incredibly vulnerable. Every coastal city on Earth is going through the same thing. It's easy to blame victims for not taking some sort of action to prevent their loss or abandon their home. I wouldn't want someone to tell me to abandon my home."

One idea Kianna is already working on is a floating greenhouse project. Using decommissioned barges, she plans to create sustainable greenhouses that function on their own. In the same vast lagoons at the center of the atolls that the US military used as bases during past wars, Kianna is working with scientists and engineers to establish a model for a sustainable Marshall Islands. "There are so many floods now on all the atolls that they're intruding on people's gardens and palm trees. How can we address this issue?"

Kianna is starting to find an answer. Using a closed-cycle greenhouse, people will have access to fresh food and water no matter the weather outside. Inside the barge, according to Kianna, "there's a fish farm below, the fish poop feeds the plants, the plants get the sweat from the glass—it literally rains inside. It's actually really amazing. Our test barge has already been in place for two years."

Sometimes being ocean people means knowing when to move on, and when to stay and fight. The type of courage Kianna and Selina embody will help inspire us all into action over the next decade.

## 2020–2021: THE SPARK, THE WILDFIRE, AND THE BACKLASH

The start of the 2020s wasn't easy. As a new decade dawned, the world was awash in thunderstorms, literal and figurative. We watched in horror as catastrophe after catastrophe materialized.

We didn't always realize it as it was happening, but we were grieving a world we knew was never coming back. We only held the promise that a better world would take its place.

Meteorologist Deke Arndt, chief of the National Oceanic and Atmospheric Administration's Climate Monitoring Branch, called the start of the decade a time of climate "goodbye"—a loss of the familiar surroundings that defined our beings, that made us who we were. Like all goodbyes, it was painful, but it enabled us to embrace our new reality. In saying goodbye, we began to recover our ability to console, to comfort, to heal, and to act.

As disaster after disaster struck, humanity faced a moment of rawness. We mixed grief, rage, and hope as we set new records for mass protest. People who just a few months before had never considered themselves to be particularly "involved" were organizing their friends and neighbors and plotting a different course. Finally, the political and the ecological began to merge. Millions of simultaneous conversations led to the same conclusion: We don't feel ready, but we have to do this. It's now or never.

In 2020, on the fiftieth anniversary of the first Earth Day, just five years after the most recent major El Niño, scientists began to receive signals that the Pacific Ocean was warming again, boosting the risk of wildfires, heat waves, droughts, floods, and tropical cyclones around the world. The El Niño conditions, combined with a bit of bad luck, set the stage for a period of global calamity outside the bounds of modern human experience.

As the US presidential election rolled on, it became clear that even the most progressive candidates still weren't prepared for the urgency with which they'd need to model an entirely different society. At times it felt like we were living in a virtual world, a caricature of all the trendy dystopian disaster movies of recent years, with events too on the nose to be believable.

Deadly heat waves occurred in major cities. Temperatures in Chicago reached 110°F (43°C) for three days straight, and the horrific toll became seared in the public consciousness—indelible images of door after door spray-painted with police markings and, in the days and weeks that followed, around-the-clock news updates chronicling a nation in mourning. Temperatures soared to similar levels in Beijing, Moscow, and Berlin—cities where such temperatures are unheard of. At the same time, severe droughts ravaged southern Europe, southern Africa, and the entire Amazonia region, while major flooding occurred throughout Southeast Asia. As a result, a global food crisis exploded, which affected a quarter of a billion people, heightening simmering international tensions. For a few scary months, the United Nations was unable to provide food aid and other relief services to dozens of countries most reliant on food imports as a result of past colonialism and distortions of capitalism—especially countries in the Caribbean, North Africa, and East Asia.

Warmer ocean temperatures in the Pacific due to El Niño set off a spate of typhoons. The worst hit China's Pearl River Delta, which includes the cities of Hong Kong, Guangzhou, Shenzhen, and Macau, with unprecedented force. Home to more than 60 million people, the Pearl River Delta recently surpassed Tokyo as the world's largest megacity. The landfall of the super typhoon, a

Category 5–equivalent that packed winds of nearly 185 miles (300 kilometers) per hour and a storm surge of up to 26 feet (8 meters), far surpassed the impact of 2018's Typhoon Mangkhut, the worst storm ever before recorded in the region. Even worse, the storm stalled out over land, lingering in the region for days and dumping nearly an entire year's worth of rainfall. Coming on the heels of the coronavirus outbreak, the storm prompted a crisis of confidence in Chinese leadership as millions of displaced people struggled to find adequate food, water, and shelter. A wave of protests spread across the country, building on those in Hong Kong in 2019, calling for more accountability from their leaders.

In the span of a few weeks, a similar-strength storm hit Mumbai—estimated to be a 1-in-650-year event, outside all historical experience. Then the long-predicted "big one" tore through Florida, carving a path of destruction from Miami to Tampa, leaving another million people homeless.

As the 2020 US presidential election neared, a rush of refugees around the world topped 100 million for the first time, tripling the number from just ten years before. Nationalist leaders in the US and Europe failed to officially recognize these refugees' legal rights to safety and instead kicked off an oppressive clampdown on migration. Echoing the words of former president George H. W. Bush at the 1992 Earth Summit in Rio, President Trump declared in a national address: "The American way of life is not up for negotiations. Period." The United States elevated its own supremacy above basic humanity. In closing borders and abandoning coastal assets, Trump's new climate policy left the majority of humanity outside the gates in a breathtaking and futile attempt at eco-fascism that drew quick condemnation around the world.

Fending off impeachment and scandal after scandal, the Trump administration used runaway climate change to justify occupying Miami's Little Haiti neighborhood—some of the highest ground in all of South Florida. With the US Armed Forces, over the course of several months, the administration's "relief efforts" worked to convert the ruins into luxury accommodations to house the affluent fleeing a destroyed Miami Beach. Meanwhile, the government did almost nothing to rebuild the rest of South Florida. Instead, in a misguided attempt to protect US interests, the administration escalated its trade war with China and dispatched the military to blockade the Strait of Malacca, effectively closing off economic relations and plunging the world economy into recession. On cable TV, pundits could only look on in horror as the president's iron-fisted climate plan of mutually assured economic destruction became transparently clear.

A group of scientists in Hawaii had predicted that this sort of thing might happen. A 2018 study examining the cumulative effects of overlapping disasters had demonstrated that climate change was increasing the frequency and severity of many types of extreme weather-related crises. Sooner or later, according to the study, our luck would run out and separate disasters would strike at once, only magnifying their impact.

"None of this happens in a vacuum," retired Navy rear admiral David Titley told me in 2014. "Climate change isn't just an environmental issue; it's a technology, water, food, energy, population issue." Like most military experts, Titley understood then as much as he does today that if we don't reorganize our society in an orderly way, conflict—rather than climate change—will compel people to abandon their homes.

"Most people out there are just trying to keep their job and provide for their family," Titley told me. "[But] if climate change is now a once-in-a-mortgage problem, and if food prices start to spike, people will pay attention. Factoring in sea-level rise, storms like Hurricane Katrina and Sandy could become not once-in-one-hundred-year events, but once-in-a-mortgage events. I lost my house in Waveland, Mississippi, during Katrina. I've experienced what that's like."

The projections have been clear for a long time: more than $1 trillion of coastal real estate in the United States is expected to be literally underwater by 2050—almost 10 percent of our current economy. That's just the coast; it doesn't factor in expanded river flood zones, areas where mega wildfires are almost certain, or farmland that will be rendered unproductive. In the midst of an impending economic collapse, investment bankers around the world didn't need an excuse to wait. At the start of the decade, as the divestment movement began to snowball, investors rushed to pull their money out of anything deemed related to the climate emergency: fossil fuel companies, utilities, insurance companies, industrial agribusiness. The revaluation of homes, businesses, and infrastructure in at-risk flood plains and coastal zones around the world kicked off a global real estate collapse in just a few months—just as Titley predicted. Shipping companies, airlines, mining outfits, automobile manufacturers—any company that relied on the continued flow of fossil fuels into the economy was at risk of bankruptcy almost overnight. The "carbon bubble," as it was called, was popping. Trillions of dollars of land and infrastructure suddenly became worthless, while distressed regions around the world abandoned their public services entirely. Mass layoffs and government-enforced austerity programs cut off resources and programs for people in need of them most. Once markets started pricing in that collapse, the insurance sued the fossil fuel industry for putting them out of business.

The global crash of stock markets was just the beginning. The recession started out the same: mass layoffs, austerity. But it quickly became clear that major changes were ahead.

On the news, climate change shifted from an occasional sound bite to days-long teach-ins, with story after story filling the airwaves. Climate change gradually became widely seen as an interconnected way to explain everything that was wrong with society. It became about people. It became about lost dreams.

Amid all the economic and climate turmoil, a sharp rise in public discontent pushed governments to the breaking point. From seemingly every direction at once, the public demanded immediate and radical change. Capitalism was imploding. A revolution was starting—a radical reimagining of what was necessary and an ambitious effort to determine what was possible in the face of global uncertainty.

"Think back to the Apollo program," Titley told me in 2014. "President Kennedy motivated us to land a man on the moon. When we talk about climate, we need to do everything we can to set the stage before the actors come on. And they may only have one chance at success. We should keep thinking: How do we maximize that chance of success?"

Against this backdrop, the climate movement morphed and linked with the ongoing protest movements in Chile, India, Hong Kong, Haiti, Ecuador, Lebanon, Catalonia, Bolivia, and Papua. Though each protest focused on different issues, from pro-democracy and anti-austerity movements to anti-war and anti-capitalism, mobilized people around the world realized that it was no longer possible, or necessary, to separate the climate emergency from their lived realities. The youth climate strikes grew into a global general strike. Hundreds of millions of people were on the streets every week, protesting the wars in the Middle East, police violence, crushing student loan debt, the collapsing economy, the lack of decent health care, and the countless other ways the future was being stolen. But the climate emergency captured the most attention.

If the food crisis was the smoldering ember that sparked a revolution in mindset, the wildfires were the blaring siren that removed all doubt of the emergency's visceral reality. In Oakland, Spokane, the Colorado Front Range, Indonesia, Italy, and Australia, images of the fires, and their aftermaths, burned in tandem with the funeral pyre for society as we knew it—charred small-town main streets and subdivisions and even a few dense urban cores—were seared into the world's psyche. Whole ecosystems were lost to the flames as the world essentially ran out of trained firefighters. The stories of survivors played in endless loops on cable television. The world felt the sheer terror of a planet that had turned against us. People watched in quiet horror as the storms, fires, floods, and droughts precipitated an intense economic recession. Direct losses from the Florida hurricane alone were more than $1 trillion, and the collapse of several multinational insurance companies threw the entire global financial system into chaos.

Since the turn of the century, an intensifying cycle of drought and hotter temperatures, combined with more and more people living in forested areas, has created the ideal environment for megafires, not only in California but also in Portugal, Greece, Tasmania, Indonesia, Siberia, and countless other places around the world. In California alone, 100 million trees died between 2010 and 2020 due to drought and invasive insects pushed into new habitats by the warming weather. And, as lightning and thunderstorms continue to spread northward with the warming weather, fires became regular occurrences in Greenland.

Because everything is connected, the warming ocean waters of the Pacific altered the flow of the jet stream, which steered precious rainfall away from Indonesia and Australia, which unleashed brutal drought and heat waves, which made fires more likely, which released more carbon emissions into the atmosphere, which surged global temperatures, which helped melt more Arctic ice, which exposed darker waters, which absorbed more heat from the sun, which warmed the oceans further. And this was just the surface-level effects. Ecosystems that had been stable for millennia before humans discovered the energy embedded in burnable dirty rocks, the same ecosystems upon which the web of life depends on the only planet where life is known to exist, were decimated in a single human life-span. Our hearts could not and cannot process grief on this scale.

In 2018, the Camp Fire almost totally erased the city of Paradise, California, killing eighty-five people in minutes—the deadliest wildfire in modern US history. Just weeks earlier, the Carr Fire had swept through Shasta and Trinity counties in California as a literal fire tornado, unlike anything any meteorologist had ever seen before. Not since San Francisco's 1906 earthquake and fires was so much lost to flames so quickly. These fires, and dozens others like them, left Californians literally sifting through the ashes of their homes and their lives.

These were not isolated incidents. Since 2003, California has endured nine of its ten biggest and most destructive fires in history. The 2018 National Climate Assessment report found that more than 50 percent of the acreage burned in California wildfires was the direct result of climate change.

The people responsible for fighting the flames came to realize that, in certain cases, fire suppression was nothing more than a symbolic act. With infernos that explode within minutes and spread at a rate faster than people can run, the first job of firefighters was getting people out of harm's way—risking lives to put out flames became secondary.

For some vulnerable regions, wildfire continued to pose an existential threat. In California, that threat risked a complete collapse of the largest private utility. As a direct result of their role in starting recent fires, PG&E declared bankruptcy in 2019. In a flailing attempt to prevent further economic harm to itself, PG&E instituted Public Safety Power Shutoffs—repeated forced blackouts for millions of people during the height of wildfire season.

What's worse, smoke from these fires likely killed at least ten times as many people as the flames from the Camp Fire—not in one year, but every year. As the deadliest consequence of burning fossil fuels, air pollution killed more than nineteen thousand people worldwide every day, and wildfire smoke made it worse. People literally breathed in the ash from their neighbors' burning homes and they died from it. Like an out-of-control fire, death became contagious.

In the 2020s, living in California came with an element of psychological dissociation. I spoke with people who decided to leave California completely. One woman, a survivor of 2015's Valley Fire, told me she was planning a move to Oregon. "We lost our home, we rebuilt ten years ago, and all of these fires now, it's like a constant terror, and I just can't live here anymore."

Lizzie Johnson, a reporter for the *San Francisco Chronicle*, wrote a book entirely filled with stories from just her first two seasons covering the fire beat. What she saw was cycle after cycle of compounding tragedies. "You [started] to wonder how long you can write about things compellingly until people just start to get empathy fatigue and stop caring," Johnson told me in 2018. "Sometimes I feel this sense of hopelessness, where I feel like my words are so limited, because all I want is for people to care and to understand what's happening."

What happened in California offered a glimpse of a horrific

future. Each new megafire brought a sense that there would never be another one this bad in our lifetimes. And then the next one was worse. A lot worse.

"I feel like people kind of need to be shaken from their stupor," Johnson told me in 2018, "and realize that climate change isn't this abstract problem that won't impact them. These wildfires are one of the first very visceral, humanistic impacts of climate change that people can identify with, in a way that they can't with melting ice caps or dying polar bears. These are little children, and parents, and seasoned firefighters dying.

"This idea that the problem seems so big that no one really thinks that anything that they can do can have that much of a difference—that's wrong. The only way we can really change this trajectory we're on is if people start caring and they start speaking up about it and pressuring their legislators to adhere to stricter policy."

* * *

In November, right before the presidential election, came the death knell for the system. A freak hurricane made landfall near Washington, DC, during an emergency session of Congress, and served as a poetic and symbolic capstone on the worst hurricane season the world had ever known.

The signal was as clear as could ever be imagined: our current system was not built to withstand the new reality we were creating. Everything, it seemed, was breaking at once.

In spite of all the chaos—or, more likely, because of it—an

upwelling of human emotion and inspired acts of defiance centered on the future. There was a tangible period of public grieving, of letting go of hopes and dreams, of preparing to abandon old ways of doing things. A new, uncomfortable, necessary path was emerging. People finally began speaking the truth we already knew: climate change impacts were beginning to unravel, reshuffle, and realign the world.

We were finally ready for transformational change. Once growth for growth's sake was seriously questioned, an opening for a new system materialized. A shift to a new method of valuation began—not on expected production or consumption but on the ability to contribute to sustaining life and civilization.

"Prepare for catastrophic success," Titley told me in 2014. "I mean, look at how quickly the gay rights conversation changed in this country. Ten years ago, it was at best a fringe thing. Nowadays, it's much more accepted. When we get focused, we can do amazing things. Unfortunately, it's usually at the last minute, usually under duress."

* * *

Success on climate change, where it can exist, will look like democracy. To build a sustainable and just world for the next century, everyone will have to participate—especially those who have been excluded from the political process for far too long. An inclusive society is a just society, in which we all listen to one another with genuine care.

Asking how the world will have to change to accommodate a

billion climate refugees by the end of the century is the wrong question. Instead, we should be thinking about how the world will have to change so no one will ever need to abandon their home in the first place.

We owe it to one another, and to all species of life, to stabilize the planet's natural systems, so that people we've never met and creatures we've never seen can grow up healthy and happy. We're doing this for all of us.

In the 2020s, all this happened much, much quicker than anyone thought was possible. As the general strikes escalated, what began as a few million students and young people following the example of Greta Thunberg blossomed into a weekly mass rally in almost every major city in the world. By the end of 2020, 300 million people took to the streets on a regular basis—4 percent of everyone on Earth—an outpouring of desire that reshaped the trajectory of human civilization. In Europe, the demonstrations routinely surpassed 10 percent of the population, effectively making normal life impossible—far surpassing Erica Chenoweth's 3.5 percent rule. Town hall meetings and citizens' assemblies throughout the world plotted new paths, neighborhood by neighborhood, city by city, state by state, and nation by nation.

Varshini Prakash's vision for the Sunrise Movement became a reality. "Because we built such an incredible coalition of people, of youth, of environmental justice and climate justice groups," she told me, "we've amassed a huge number of ordinary citizens and Americans who see themselves reflected in the Green New Deal and see the way in which it can directly impact and benefit their lives."

After we ushered in a new generation of politicians in 2020, they emerged to fight for people first, not for money or power. Spurred on by an engaged citizenry fed up with the country's leadership, this emboldened group of new politicians countered the global rise of authoritarian strongmen concerned only with their own power and started to do the hard work required to deal with the climate emergency.

"For the first time," Prakash told me, "we had a window of opportunity to pass the kind of policies that we need[ed] to pass to stop the climate crisis. A chorus of people all around the world called on their own nations to protect our future."

When I spoke to her before the election, she had told me, as a part of her vision for the future, "I'm imagining right after the election, November 2020, there's a fire season that's continuing to happen and quite literally lines of people are bringing the ashes of burned-down houses and trees and their homes, and dropping them on the doorstep[s] of our politicians by the hundreds. It became a thing that's not about activists. There's no ability to call the people who are taking action 'activists' because there's so many of them that they just become people calling for something that's going to save our society. That's the level of scale that I want to reach."

The emergence of a new political consciousness dovetailed with a growing sentiment in popular culture, as celebrities and artists turned their creative energy and influence toward raising awareness about the climate emergency and galvanizing public support for radical change. Perhaps the most important development was the rise of a popular reality TV show in the United

States that asked a simple question: What are we going to do? It was the question everyone had on their minds, of course, and the show's producers created an ingenious way of turning everyone's shared existential crisis into a path forward. By randomly assembling one hundred people—a public citizens' assembly—the show's cast resembled America. Together, they built up a vision of what they wanted the world to look like in 2050, then worked backward. With the help of dozens of experts, they settled on a pathway for change that the US could implement right away. They built a hopeful vision of a future that works for everyone, along with a plan to make it happen that everyone could support. Their plan was much more radical than most politicians were willing to admit.

They announced a goal for the country's energy use to be 100 percent carbon-free by the end of the decade. They called for public ownership of all utilities and an immediate end to all fossil fuel subsidies.

They asked their representatives to invest in rural regions and establish a fully funded national institute to study regenerative agriculture. They also asked their representatives to sign into law a national car buyback program combined with comprehensive city redesigns that aimed to completely eliminate cars by 2040.

In addition to a four-day workweek, they demanded universal guarantees for housing, health care, and employment—all of which would help transform the current economic system into a completely circular economy by 2050.

Finally, they demanded recognition of Indigenous sovereignty and the establishment of a permanent fund for climate reparations.

To pay for these radical ideas, they called for a wealth tax on billionaires. In the US, the plan went even further than Bernie Sanders's Green New Deal, which was by far the most ambitious plan ever proposed by a presidential candidate. And best of all, economists of all political leanings agreed the plan would pay for itself by 2030.

In January 2021, a new president was sworn in with climate change as a day-one priority. By the end of the first week, Congress got to work on passing most of the ideas the citizens' assembly had come up with. A new era of history had begun.

## 2022–2023: IMAGINING A NEW SOCIAL COMPACT FOR A RAPIDLY CHANGING WORLD

To achieve our dreams, we needed democracy to work better.

In the United States, a series of structural changes in 2021 helped pave the way for the rapid transformation in all aspects of US society. Congress banned all fossil fuel industry advertisements, expanded the Supreme Court and instituted term limits on it, and abolished the filibuster in the Senate.

Washington, DC, finally became a state, and US-controlled island territories were also granted statehood, in recognition of the existential threat climate change posed to them.

Puerto Rico became the fifty-second state, the Virgin Islands, the fifty-third. Soon thereafter, Guam and Northern Mariana Islands were declared the fifty-fourth and fifty-fifth states of the Union, followed by American Samoa.

At a ceremony unveiling the new US flag, with fifty-six stars, the new US president said it represented the unity of the American people and our striving for one goal: our shared survival and hope for a brighter future for everyone.

Perhaps most important, recognizing the duty to the tribal nations that far predated the country allowed for a new chapter of dialogue in an attempt to rebuild trust and share the lands we occupied.

Kelsey Leonard, an enrolled citizen of the Shinnecock Indian Nation and tribal co-lead on the Mid-Atlantic Regional Planning Body of the US National Ocean Council, told me that she believed that this essential step would change everything, not only for Indigenous nations like hers but also as a lasting symbol of what it means to work toward a consensual relationship between one another and with the planet we all call home. "The principles of right to self-determination and free prior and informed consent are enshrined within the United Nations Declaration on the Rights of Indigenous Peoples," she told me. "Decolonization means much more than just asking permission to build a piece of infrastructure once you're already halfway through the planning process, or making a land acknowledgment. It means stepping aside in real and measurable ways."

The traditional lands of the Shinnecock Indian Nation are on Long Island, where in recent years developers have built wind turbines despite a minimum of direct dialogue with the tribe. It's the kind of process that, over time, destroys relationships.

"From my own experience at Shinnecock," Leonard told me, "we've had to deal with quite a lot of wind energy proliferation and the lack of consultation that goes into that." In this case, Leonard said, renewable energy development replicates patterns of colonialism. "The ways current wind proliferation, wind permitting, occurs sits outside of the control of Indigenous nations in the US. It's been very much a process that's been controlled and dominated by the state and the federal government[s]. They jockey between the line that is state waters and federal waters and out to the Exclusive Economic Zone. All of that happens under the auspices that Indigenous waters don't exist. They just ignore the fact that there was never any ceding of water rights to the states or the federal government, and then the permitting processes occur without our consent."

Leonard knows that the prevailing attitude toward renewable energy development has heightened urgency in the middle of our climate crisis. Asking developers to include her people from the very beginning of the process, though, is an important step toward bringing about a more just and sustainable world. "I don't think tribes are anti-wind. It's how the process is currently conducted, in the sense that the placement of turbines doesn't actually consider our fishing territories or canoe routes, any aspect of our political or cultural existence that has to do with the area that's being now occupied by turbines."

These are the kinds of disputes that don't happen in a culture built on care and consent. As the world entered an era of radical change, we learned how to simultaneously do the slow work of building relationships with one another.

* * *

Imagine it. What would it be like if every family, every neighborhood, every farm, every city, every nation on Earth were all working toward a better world?

In Puerto Rico, the change has been revolutionary. Hurricane Maria was the worst disaster in Puerto Rico's history—a superstorm that plunged the island into a humanitarian crisis. In 2022, five years after the storm, a transformational change was slowly realized. Statehood—a controversial step—has allowed Puerto Ricans a voice in their own future, and resources to rebuild in a way that started to attract members of the diaspora back home.

Marisol LeBrón, a professor at the University of Texas at Austin, told me that both political and economic sovereignty were necessary for Puerto Rico to truly flourish. "The issue throughout the Caribbean and Latin America is that political sovereignty without economic sovereignty just creates new forms of imperial domination," she said. "Without economic sovereignty, political sovereignty is essentially meaningless. A real decolonization would mean that Puerto Rico would be given reparations for over a century of colonization by the US and for centuries of colonization by Spain. Also, reparations for the centuries of extraction and exploitation that have happened. Without that, Puerto Rico will be completely unable to start to rebuild in a way that promotes equality and justice for the people living there.

"One of the things you saw a lot after the storm were these kind of small-scale collectives coming up that sidestepped the state completely. The people providing for the people. And addressing the needs of the people. Those efforts have been happening for the past ten years, and are just gaining a lot more traction as people see the situation and the state, the way it was, was not going to provide for them or keep them safe in the face of these ongoing crises."

Debt forgiveness, as well as an overdue plan to fund the rebuilding of homes and infrastructure destroyed during Maria, made it possible for Puerto Ricans to imagine a truly hopeful future, one that finally shook free its colonial past.

* * *

Meanwhile, in 2023, world leaders gathered again in Europe for a global reassessment of progress since the Paris Agreement was signed. The conference was filled with tension and open conflict about the paths societies were following amid rapid change.

Despite apparent progress in the US, the world needed to strengthen the commitments made in Paris, which had been woefully insufficient. Each country agreed to an emergency review before a new global climate summit in 2025.

Toward the end of the two weeks of sessions, the US delegation splintered. The fossil fuel industry still had a lot of power, though it was rapidly dwindling, as its actions began to be widely seen as immoral and even illegal. For China, the Belt and Road

Initiative—its effort at spreading power and influence across most of the world—became a key negotiating point. The African Union and European Union held marathon talks about reparations for colonialism.

It's true that the US had come a long way since the late 2010s, but other countries were rightly skeptical. At last, a breakthrough: the Marshall Islands helped to broker international commitment to a Global Marshall Plan, directed by developing countries and funded proportionally by every nation in the world on the basis of historical greenhouse gas emissions, a multitrillion-dollar, twenty-five-year effort toward a rebalancing of humanity.

The Global Marshall Plan was the most ambitious initiative humanity had ever undertaken. Its main goal was not economic stability or even emissions reductions, but a quarter-century effort to gift a thriving planet to future generations. It was intentionally vague—the full text of the agreement was only eight pages long—but its power was in setting up a system of binding dialogues within and between every country on Earth. At the community level in every city of every size worldwide, discussions would take place in an attempt to create locally determined paths forward that were consistent with the 1.5-degree Celsius goal set in Paris. Finally, we were able to decide our future by ourselves, with one another.

As the plans started rolling in, it was clear that what was happening was far beyond even the dreams of the Marshallese organizers. Trillions of dollars were spent on renewable energy and agricultural research, tens of thousands of miles of high-speed rail lines were constructed, and seeds were distributed to backyard farmers and subsistence agriculturalists. There was also support for teachers, doctors, and artists. People were ready for revolutionary change.

## 2024–2029: THE GREEN NEW DEAL GOES GLOBAL

By the mid-2020s, a global movement took hold that centered around the difficult necessity of reimagining our relationships to one another and the world. A series of legal victories brought the authority that was necessary to reshape the playing field, and fast.

How quickly could the old rules change? Could an international court mandate intergenerational equity? Could there be criminal charges filed against the CEOs of fossil fuel companies, holding them legally responsible for crimes against humanity?

In the waning days of 2019, a breakthrough changed the playing field overnight: the Netherlands Supreme Court ruled that Urgenda, a Dutch environmental group, was right to claim that a failure to address climate change violated human rights. Around the world, from Peru to Canada, dozens of related cases opened different legal pathways to ambitious climate action. By the late 2020s, the verdicts put a positive mandate on countries to revolutionize their economies and deliver a real guarantee of a habitable planet.

*Juliana v. United States*, in particular, had a plot suitable for a Disney movie: an eclectic group of twenty-one kids (and their lawyers) filed a lawsuit in 2015 arguing that the federal government's lax climate policies had violated their constitutional rights to life and liberty and should adopt a science-based plan to reduce emissions.

At the heart of this lawsuit was the principle of intergenerational equity. In essence, the twenty-one plaintiffs in *Juliana* said that the federal government's refusal to take serious action against climate change unlawfully puts the well-being of current generations ahead of future generations. It's a horrific injustice that children have to grow up wondering whether the planet they live on is going to quickly become incapable of supporting life in all its beautiful forms.

In 2020, however, *Juliana v. United States* was dismissed by the Ninth Circuit Court of Appeals. In her strongly worded dissent, US District Court Judge Josephine Staton wrote: "In these proceedings, the government accepts as fact that the United States has reached a tipping point crying out for a concerted response—yet presses ahead toward calamity. It is as if an asteroid were barreling toward Earth and the government decided to shut down our only defenses. Seeking to quash this suit, the government bluntly insists that it has the absolute and unreviewable power to destroy the Nation . . . considering plaintiffs seek no less than to forestall the Nation's demise, even a partial and temporary reprieve would constitute meaningful redress. Such relief, much like the desegregation orders and statewide prison injunctions the Supreme Court has sanctioned, would vindicate plaintiffs' constitutional rights."

In 2024, the youth refiled their case with a slightly different legal argument: They wanted to make burning fossil fuels against the law in the United States, a violation of the youth's Constitutional rights, as of January 1, 2030. Doing so would enshrine the past four years of rapid progress toward a Green New Deal into the most primary legal document governing the country. Against all odds, the youth won. In the Supreme Court's decision, they cited the Urgenda case, forever linking climate action and civil rights in American law. A period of rapid decarbonization began at once across the entire US economy.

Michael Burger, executive director of the Sabin Center for Climate Change Law at Columbia University, predicted something like this could happen when I spoke with him in 2019. "The idea that the International Criminal Court is going to step in seems kind of far-fetched, obviously. The idea that individuals will be held personally, criminally liable for corporate actions over the scope of . . . industrial and post-industrial human civilization is, it sounds, unlikely. But if there are discrete acts, maybe there's something there. Like, if it has become a crime to emit greenhouse gases. The thing about criminal prosecutions is that the crime has to be written out and clearly articulated. So you'd have to create those crimes."

More likely, he said, is the United States would establish a new department in the executive branch whose job it is to advocate explicitly on behalf of future generations. That's exactly the kind of thing the original Juliana case called for, but they got even more than that. The Supreme Court's 2024 ruling unleashed a wave of legislation and executive action that solidified America's responsibility to repair the climate damage it had caused worldwide for centuries.

"Right now, we're at a point in time where some of these cases, like . . . the Urgenda case in the Netherlands, and potentially some other cases, are establishing that these domestic constitutions or regional human rights regimes protect individuals from climate change and obligate governments to do more than what they're doing. So that could be running to courts all around the world and saying, 'Your failure to have a climate plan in place, adequate to reach the 1.5-degree target, is a violation of our constitutional right to life.' And courts [are] saying, 'Yes, that's right. You have to do more.' So that seems like a good possibility to me."

As court systems around the world began to swing to the side of youth and future generations, all bets were off in terms of how rapidly climate action could begin to proceed.

In India, animals, birds, and rivers were granted legal personhood status, and a court ruling found that climate change harms there must be limited to protect them. In the Cook Islands, courts ruled that the Pacific Ocean similarly deserved protection equal to humans. Finally, the Earth had legal protection that justified its irreplaceable value.

* * *

By 2030, we may find it difficult to remember what life in the 2010s was like.

The period from 2020 to 2030 will be both a truly terrifying and a golden era in humanity. Nothing like it has ever occurred before. With any luck, nothing like it will ever be necessary again. We will come together—some people willingly, some because they will run out of options or excuses—to make the best decisions in the best interests of the planet. We will realize that by showing people, who are desperate for solutions, what is possible, what we can do, we might ensure that future generations are handed a world that isn't on the verge of going to hell.

In the US, if we do what we need to, net emissions in 2030 might be just 10 percent of what they were in 2020. Globally, emissions might fall by 40 percent. We will emerge from the liminal space of climate catastrophe into a still-uncertain world, but one that is bending back toward life.

# 2030–2040: RADICAL STEWARDSHIP

*We live in capitalism. Its power seems inescapable. But then, so did the divine right of kings. Any human power can be resisted and changed by human beings. . . . The name of our beautiful reward isn't profit. Its name is freedom.*

—Ursula K. Le Guin

Designing and executing a plan to radically reduce emissions is where most discourse on climate action stops. But it is really just the beginning.

We need more than just renewable energy. We need more than just tearing down the fossil fuel industry and capitalism. We need to develop a whole new type of human society.

If the kind of thing we're most focused on is replacing gas-powered cars with battery-powered ones, we'll have missed the whole point, and we'll be well on our way toward re-creating the same system that got us into this mess in the first place. In this moment of transformational change, we need to start by asking foundational questions, like, What is a good life, and how can it be possible for everyone? We know what needs to change is almost everything that makes up society as it is today: the systems of buildings, transportation, and energy that make up our cities and towns, but also our democracy, our justice system, and the way we value one another and ourselves.

I have no idea what this will look like, but I know how to find out: it's as easy as listening. That's how a new politics, a new way of relating to one another, will come into being. That's the difference between renewable *energy* and a renewable *economy*.

The 2030s—the dawn of the era after we've reached peak global emissions—will be just as pivotal for the fate of civilization as today's efforts are to radically and urgently change course. Long before the tar sands, the urban sprawl, the industrial agriculture become relics of the past, we'll need to begin putting together a framework for what will come next.

Revolutionary ways of looking at the world—such as the design concept of biomimicry, which entails closely examining the ways nature works—will move us from our current systems to systems that truly work for everyone. For the vast majority of people on Earth, it's likely that life in a sustainable economy will be much richer, more pleasurable, and more satisfying than life under capitalism.

This is a radical concept for many people. Life in the 2030s will look and feel different than life does now under late capitalism, when egregious inequality, racism, and poverty have made it transparently obvious that the system is designed to work for a specific few—and probably not you. Supplanting the current system will be a difficult struggle. We're up against the richest and most powerful industries in the history of the world, but we've got life itself on our side.

After a revolutionary change in mindset in the 2020s, we'll be ready to get to the hard work of making fossil fuels obsolete and expunging Earth-intensive ways of living in favor of systems that are designed to put our personal and planetary health first.

With America's lead, and a spirit of justice, it's possible to almost entirely eliminate fossil fuels in the near term. The United States peaked its emissions in 2005, during George W. Bush's second term as president, pushed mostly by a switch from coal to natural gas. By 2035, the United States could have an entirely carbon-neutral economy, with a switch from natural gas to renewable energy. China, the world's largest producer of greenhouse gas emissions, is on pace to peak emissions by around 2030 and could begin a sharp decline soon after. This, or something like it, is the timescale that science says is necessary to avoid a climate system that is warming beyond our control.

That timescale includes not only completely switching to carbon-free electricity generation through a blend of wind, solar, geothermal, and hydro power. It's not even only phasing out petroleum-powered engines from all cars, trucks, trains, ships, and aircraft. It's also doing the hard work that not many people are talking about right now, like finding alternatives to fossil-fuel-based chemical fertilizers in agriculture and finding new processes for cement production. It's a remaking of our entire economy on an emergency timescale.

This kind of transition is not possible without leaving capitalism behind. The perpetual-growth model is simply not built for an era of rapid planetary change. In a world where the richest 85 people in the world own as much wealth as the bottom 3.5 billion, and the wealthiest 10 percent produce 49 percent of all emissions, it's not individual choices that are driving climate change. When we realize that rich people have stolen our planet's habitability for themselves, we will demand revolutionary change. Systematically shutting down the fossil fuel industry while simultaneously building a decentralized energy system is inherently anti-capitalist, but it's just a first step. For social and economic liberation in an era of climate catastrophe, we've got to go much further.

## LIFE AFTER CAPITALISM

If what you've been doing for hundreds of years has brought you to the brink of a mass extinction, maybe it's time to try something new.

To realize the future we want, we need to consider options beyond our current system. Kate Raworth thinks she has the start of an answer. In her 2017 book *Doughnut Economics*, she reimagines what a global economy could look like in the twenty-first century. A circular economy, as she describes it, simultaneously prioritizes the well-being of humanity and respects our planetary boundaries.

Raworth's doughnut chart helps us visualize a safe operating zone for a thriving human society on a planet with finite resources. The inner circle features equity and justice, among other necessities—nonnegotiable indicators of a healthy civilization. The outer ring features environmental indicators that, if left unconstrained, threaten our existence. Her idea is simple and revolutionary: keep society within the doughnut, the safe middle zone that will result in a flourishing world.

"In a circular, or cyclical, economy, there's no such thing as waste," Raworth told me. "It's just a resource in a wrong place. That applies to carbon dioxide as much as it does to throwaway plastics. That waste from one process is food for the next."

Think of it as recycling, but on steroids. Not only will we reuse everything; we'll also transform all our manufacturing and industrial processes so that the concept of waste itself is obsolete. For us to get to where we need to go, we need to become a society with a cultural focus on repair and maintenance, rather than innovation and efficiency.

"Today's linear and degenerative industrial system is the take-make-use-lose system: we take those materials, make them into stuff we want, use them for a while, and throw them away." A society that eliminates the concept of waste, Raworth told me, will no longer treat its members as disposable.

Although Raworth's presentation of this information is fresh and exciting, living within our means is definitely not a new concept. A circular economy goes by many names—"regenerative," "degrowth"—a specific application of sustainability that focuses on meeting needs, rather than keeping up with our wants. Unlike post-scarcity economic theories and philosophies—like transhumanism or eco-modernism or Fully Automated Luxury Communism, in which we scale up automation to reap the benefits of record worker productivity—a circular economy de-emphasizes growth for growth's sake. Because scarcity does exist, especially on the timescale that is relevant for rapidly transforming society in an all-out response to climate change, the push for a better world must have the principles of equity and justice at its core. Acknowledging that every living thing matters, and that right now life itself is in peril, Raworth's circular economy minimizes waste to maximize the ability of the system to care for everyone.

Blanketing the desert with solar panels or filling the roads with electric cars is no more a sustainable solution than monoculture agriculture is a solution to world hunger. Nor does a carbon tax have the ability, on its own, to reform capitalism. Such market-based mechanisms simply reinforce the status quo that got us into this mess, and trust in the flawed logic that more is better. Degrowth does not mean austerity. It's trusting that the only planet in the known universe that's capable of harboring life will provide more than enough for all of us if we simply take care of it.

Raworth said that her concept of the doughnut has helped frame the kinds of solutions that need to happen. And, even more than that, it could be a practical application of a change in the operating philosophy of human civilization itself. We'd be moving away from the idea of nature as an expendable resource to be used up and toward an idea that fits more with an ecological understanding of humans as just one part of an interconnected system. A circular economy, deeply rooted in justice and equality, is what could bring everything back into balance.

The economic system that will emerge in the 2030s will start with the doughnut, but the system could end up unlike anything anyone is considering right now. A few key ideas that are already coming into focus could help form core parts of how society would function. Our labor, everything we create, and our relationships will wind up embedded in a new reality.

Because the word "capitalism" carries with it unnecessary baggage, Raworth believes it should be discarded entirely, not just tweaked. Instead of replacing it with "socialism" or "communism" or another loaded term, she told me, "let's just rethink thriving and then ask ourselves what kind of economy would result. We would choose to drop the old '-isms' of the past."

No matter what we call it, or exactly what policies it contains, what our next economy will look and feel like is already coming into focus.

## WHAT WOULD IT FEEL LIKE TO TRANSITION TO A DEGROWTH ECONOMY?

Putting radical ideas like these into action may seem scary, from the vantage point of today. In order for them to catch on quickly, the ideas have to be unambiguous as well as aspirational. With a shared civilizational goal of improving lives instead of making rich people even richer, we'll have all the moral clarity we need for this new phase of history.

Uganda could become a test bed for pioneering new kinds of agricultural methods that prioritize species diversity and perennial crops. Factories there could begin producing plant-based plastic alternatives, which would then be exported throughout the continent on a newly constructed high-speed rail system.

In Iran, the focus could shift away from fossil fuel production toward carbon-negative cement, with universities there specializing in developing new architectural styles that would enable a revolution of building practices around the world.

The Caribbean islands could shift away from highly volatile and exploitative tourism-based economies toward a broader philosophy of coastal living, with strong community ties and a spirit of resilience. They could help pioneer new forms of wave energy that might power every coastal city on Earth with little environmental impact.

In the US, the old Detroit factories could be put into production again, as the center of the century's green economy. Almost everything that powers the transition to a carbon-free economy could be produced there, in at least some quantity, from railcars bound for Uganda to electric water heaters shipped across town. Detroit's population might double, even surpass its all-time high when its economy was geared around gas-guzzlers.

In small towns and cities around the world, from India to Peru to Egypt, excitement would be in the air as humanity finds common purpose in working to avert catastrophe by building the kind of society we have deserved all along.

* * *

Here's what Raworth imagines would be necessary for these kinds of visions to become a reality:

> ***By 2040, we started to move back within the sphere. We're not going to have eliminated all human***

*deprivation, and we're not going to be back under 350 parts per million. But we started to significantly bring ourselves back within that space. Then the question is, what are the economic dynamics that are going to be in place?*

*First of all, we're in a world where governments are no longer standing up and giving a state-of-the-nation kind of report, focusing on GDP and its incremental tiny changes . . . Is it 2.0 or 2.3 percent this quarter? That's gone. We have governments that are required to report how we're doing in terms of improving the extent to which we meet all the social needs of all people. The policies we've got in place are starting to bring us back within all those different planetary boundaries. So it's not just about climate change. We talk about a thriving planet, and we have a more informed public and government and business [sector] that can talk about these different planetary boundaries.*

*We've become carbon-footprint literate in the past few years. But by 2040, we will be literate across this diversity of planetary boundary footprints. . . . This doughnut image, or something very like it, will be the metric of success.*

*We'll be able to see a much closer linkage between today's policies and the future. There'll be a lot more social and ecological accountability in governments. . . .*

*We won't talk about growth. We'll talk about thriving. We'll have shifted the fundamental metaphor of what we think economic progress is.*

In order to begin a rapid transition to a society that works for everyone amid a worsening climate catastrophe, we've got to imagine the alternative as more appealing—even for the capitalists!—than our current system.

The goal of liberation from climate change, after all, is not only surviving but thriving. Adopting the Global Marshall Plan could help balance the global economy to the point that the gap between rich and poor countries will begin to narrow quickly, bringing billions of new voices and hopes and dreams into conversation with one another. Transformative knowledge would come from everywhere at once, and a diversity of opinions, food, clothing, memes, and culture would give a richness to our existence unlike any time before in human history. This is what liberation would feel like.

According to Raworth, part of the engine of this transfer of imaginative vision will be the radical democratization of the means of production:

*There's an unprecedented opportunity for technologies to go deeper and say, "We don't just want to redistribute incomes. We want to pre-distribute the sources of wealth creation, pre-distribute access to create and share and build our knowledge."*

*Rather than centralizing the returns to the one percent, it creates a more distributive flow through the economy so that you get more health and balance in terms of smaller scale ownership of enterprises and reinvestment in local economies. Global sharing of ideas, but local production.*

Raworth calls this a "stewardship economy," a shift in mindset from producer–consumer to co-creation and sharing. This is the inherent optimism of a degrowth economy. That radical change will require us to establish and maintain relationships with one another like never before. We'll embrace cooperation instead of competition because we'll see how much better off we all will be in a world that works together as a system.

For places like Puerto Rico, India, and Uganda, the centuries-long effects of colonization will start to be reversed, proactively, through political and economic self-determination, reparations, and aggressive climate action.

* * *

As we consider life beyond capitalism, we recognize ours is a moment of simultaneous destruction and creation, and that's necessary in order to identify the rotten core ideas and concepts of the old paradigm we've been operating under for hundreds of years: the colonialist, extractive, white-male-dominated society that works for only a certain type of person at the expense of literally everyone else, plus the planet's ecosystems. We should be in the process of actively destroying that paradigm and coming up with an entirely different system.

A relational society built on trust and consent rather than on domination won't just happen on its own; it has to be fostered. Raworth has talked about twenty-first-century economists as gardeners rather than spreadsheet optimizers. All of us should get a stake in what the key issues are in the new society, not just those with the most power, and that requires a dialogue-based world that's constantly regenerative and, as a core principle, nurtures all life. Let's imagine what this could look like.

## 2030–2032: DIRECT DEMOCRACY FOR THE CLIMATE AGE

The early 2030s demanded big decisions, which the people of the world had to make very quickly. Combined, the urgency of the moment and the magnitude of our previous actions required us to reimagine how we practiced democracy in a globally connected world.

On a scale we never thought possible, decentralized decision-making and direct democracy became the new reality. Following the victories of the community-level decision-making of the Global Marshall Plan in the late 2020s, a new spirit of optimism spread around the world. Leaders everywhere started doing what was necessary, not just what was possible. Together, we overcame the dangers of short-term thinking that capitalism helped foster in the twentieth century and that caused a breakdown in our ability to make any meaningful decisions on climate change.

We began to see democracy not as a tool to rubber-stamp the wishes of the rich and powerful but as an engine for self-determination all the way down to the local scale, via a flourishing culture of citizens' assemblies. In the early 2030s, some of the most successful citizens' assemblies were in India, the emerging epicenter of both climate disaster and climate solutions. For India, the world's largest democracy, this time involved taking a necessary, albeit complicated, first step toward ensuring everyone's voice mattered.

In thousands of meeting halls across the US, too, people gathered to come into direct agreement with one another on the details of the transitions unfolding. Slowly, but with growing confidence, we broke free from centralized pockets of entrenched power and profit, which allowed us to decentralize and redistribute means of production to foster more just representation in the country and eventually in the economic systems around the world. Embracing a regenerative economy empowered us to reclaim the climate and, in the process, restore justice.

In this decade, as the United States, China, and countries throughout Europe reduced emissions, India—now the world's most populous country—emerged as the largest emitter of greenhouse gases. By every measure, India was more vulnerable than ever to the worsening extreme weather.

As the century's fourth decade unfolded, India's prolonged heat waves grew much worse and ranked among the deadliest extreme weather events on the planet. In Kolkata, these heat waves occurred every year, while residents of Mumbai, Delhi, and Chennai regularly endured agonizing stretches of extreme heat too. All told, more than 800 million people were exposed to potentially lethal temperatures in India as a part of their regular, day-to-day routine.

While India stared down the increasingly likely specter of heat index values approaching 170°F (77°C) by the end of the century—temperatures that are physically not survivable to humans and hotter than anything experienced on Earth in at least 50 million years—its citizens mobilized and declared, as one people, that air-conditioning was a basic human right, bringing artificial cooling to the 300 million Indians living without it, along with the right to electricity, health care, housing, and a universal basic income.

With its shift toward direct democracy and a comprehensive set of social safety nets, India emerged as the moral center of global geopolitics, offering countries around the world a working model for how to enact policies that center previously marginalized communities and, as a consequence, produce both a more regenerative economy and a more egalitarian society. Every decision creates winners and losers, and finally the huge number of people systematically shut out of a good life for centuries began to win.

Across Africa, countries followed India's lead and implemented similar policies to ensure that the remaining 600 million people without electricity on the continent were finally connected to the rest of the world. Long a victim of colonialism and its extractive and exploitive practices, Africa emerged in the 2030s as a model of peace, prosperity, and promise. By the end of the decade, most countries in Africa were considered middle-income.

Similarly, twenty years after Hurricane Maria, Puerto Rico started to benefit from the very forms of wealth and resources that had been stolen from them. The work the environmental justice movement in Puerto Rico did in the 2020s following Hurricane Maria, running food banks and rehabilitating abandoned buildings, reclaiming them as community spaces, paid off in the form of flourishing local economies.

"In the aftermath of the storm," according to professor Marisol LeBrón, "people organized. And people were saying, 'Whether the government, the federal or the local government, does anything, we're going to try and do what we need to do to protect ourselves and make our communities resilient and safe.' People [began thinking] about new ways to do that."

Throughout Puerto Rico, grassroots initiatives dealt with the problems most associated with climate change, with the critical social and environmental vulnerabilities exacerbated by Hurricane Maria. "People started trying to take these things into their own hands at the community level," LeBrón told me, "trying to do small-scale work, the slower and the harder work of trying to get people to know more about the environment and their impact on it." Residents of Puerto Rico understood that it was up to them to reclaim the landscape and fight for energy independence, because the next climate-related catastrophe could cripple them at any moment. At the same time, they pushed back against neoliberalism and its assumption that citizens can and should get by with less assistance from the government. They demanded, en masse, that the state continue supplying local communities with the resources and services they needed to recover and then thrive. This demand for a reciprocal partnership was at the heart of the long-delayed trend of decolonization that occurred throughout the 2030s around the world, from Puerto Rico to India to Africa. In formerly colonized countries, like Canada, the United States, and Mexico, governments began to cede land back to the Indigenous nations it was stolen from.

## 2033–2038: BUILDING A NEW WORLD

Across the developed and developing world in the 2030s, rapid and lasting change occurred in every aspect of society. Because that's what the IPCC had said was necessary in 2018, that's exactly what we did.

In the United States, we realized that we preferred spending time with one another rather than maintaining our stuff, so the default lifestyle of a single-family home in a car-based neighborhood started to become obsolete. Voting in a million city council and regional planning meetings around the country, people agreed to rezone their neighborhoods. Duplexes and triplexes became the new default dream, with more and more people living next door to friends and family instead of across town or across the country. Massive investments in public transit and bike infrastructure made travel cheap, safe, and efficient. Small businesses and corner stores once again flourished.

Across the developing world, worsening climate impacts and the lingering air, water, soil, and noise pollution of a fossil-fuel-based economy convinced governments to prioritize public health and well-being as a central focus of how they spend the massive amounts of reparation funds from the companies and countries that created the crisis, money that came after successful legal challenges upheld and enforced the agreements reached in the UN summits of the 2020s. The construction of new hospitals, schools, libraries, parks, cultural centers—the basic foundations of what makes life worth living—leveled the playing field for people who for centuries had been on the short end of the global economy. In most places around the world, the day-to-day experience of adapting to climate change meant more people were able to survive and thrive in ways that had never been possible under capitalism. Complemented by a billion new jobs around the world, tens of trillions of dollars were invested in local communities with a single goal: to rebuild human civilization in a way that works for everyone.

* * *

By 2030, two-thirds of the world's population was living in cities. This required a new approach.

My city, Saint Paul, Minnesota, put their plan together in the 2010s and prepared for a radical transformation by 2040. I saw this transformation occur all around me. Saint Paul describes itself as "the most livable city in America," but the weather here rapidly became more extreme, just like everywhere else. By the end of the century, if nothing changes, the number of very hot days will triple and rainstorms will be about 50 percent heavier.

For my city to complete its radical transformation, it began implementing steps to

- electrify and remodel every building in the city to exclusively use renewable energy;
- install community-owned solar gardens;
- reduce driving 40 percent by eliminating parking, adding congestion pricing, and removing highways;
- allow self-driving electric cars, e-bikes, and scooters for last-mile transit;
- add hundreds of miles of new bike lanes and sidewalks on every street; and
- reduce waste by 80 percent.

Comparable in scope and ambition to the plans in New York City and the cities in California, the initiative in Saint Paul—in the heart of the Midwest—demonstrated to the rest of the country that carbon-free cities were possible. Its success triggered similar initiatives here and abroad. Just across the river, Minneapolis moved forward with a near-term vision for a total overhaul—the first major city in the country to approve the elimination of all single-family zoning.

A new streetcar route started passing through our neighborhood, Highland Park, in 2033 when my boys were in high school.

An old Ford assembly plant down the road was transformed into an entirely new walkable, bike-friendly, zero-carbon neighborhood, a construction project that was completed in the 2040s.

As Saint Paul's vision of a car-free city became a reality, no one in my city lived more than a quarter mile from a transit hub, where they could charge their electric scooters, catch the bus, and never have to fight traffic again—a transit utopia.

In the early 2030s, as the US began implementing an expanded Green New Deal, we started to focus on structural ways to build empathy and closeness that would help communities be more resilient in an era of constant change.

In his first speech as Saint Paul's first African American mayor in 2018, Melvin Carter spoke of his unlikely role in connecting the city's past and future. Rondo, the neighborhood Carter grew up in—and which included his grandparents' house—was partially demolished in the 1950s to make way for the construction of Interstate 94. The neighborhood never quite recovered, but Carter envisioned a better future. "We're afraid equity means turning our back on those who are doing well," said Carter in his inaugural speech, "that bikes, density, and transit will change our neighborhoods. . . . We're afraid of having to choose—between preserving the traditions our parents loved and building the city our children deserve. I understand those fears. That freeway cost my family everything, so I'll be first to admit change can be scary."

Rondo had a more than hundred-year history as a thriving center of Black culture. The original Federal Housing Administration of 1934 established discriminatory home-lending guidelines that drew literal red lines on city maps across the country, including in Rondo, mostly in Black and low-income neighborhoods that were deemed to be a high risk to banks. The impact of redlining persisted for decades in Saint Paul, like in most other cities, and gave the impression that these neighborhoods were expendable—sacrificed for infrastructure to benefit, almost exclusively, a white majority. For too long low-income neighborhoods lacked adequate sidewalks and pedestrian-friendly infrastructure, partly because cities converted them into roads for wealthier people traveling from the suburbs to drive on. Consequently, people of color remained skeptical of large-scale infrastructure projects put forward by white environmentalists. But as these projects started to benefit everyone in the 2030s, regardless of race or socioeconomic background, the public's perception of them started to shift.

The Green New Deal included proactive policies on equality and justice to repair centuries of discrimination and racism—not only in Saint Paul but in cities and communities throughout the Midwest and up and down the East and West coasts.

In 2033, a freeway removal project in Carter's climate plan helped restore some of what was taken away by the redlining and car culture of the late twentieth century. A new Rondo was born as a tangible reminder that it's never too late to correct our past mistakes and move forward, together. Along with his plan for removing car infrastructure, Carter received federal funding to rebuild the neighborhoods that were bulldozed decades ago with a new universal housing guarantee. Saint Paul residents were hired to do this work, and countless other jobs, with a full social safety net—including free health care, free college, and a universal basic income that would provide every person in the US with the basic means of survival so they could spend their time pursuing their dreams. We started working toward restoring not only stolen Black neighborhoods but also land stolen from Indigenous peoples hundreds of years ago. By 2040, St. Paul served as a model city that works for everyone.

"We all exist in a long line of immigrants and refugees who've conquered incredible odds to find hope in St. Paul," said Carter back in 2018. "This beautiful, diverse city they built for us is our ticket to the future."

This is what decolonization and transformative change looked like in practice during the 2030s.

* * *

On the West Coast, as our age of megafires continued, people were faced with a tough decision: Should they stay or should they go? Changing patterns of rainfall and heat shifted local ecosystems north at an alarming speed—dozens of times faster than scientists had anticipated. Without a radical shift in how and where they built houses, Californians would have lost more and more towns and neighborhoods at the edges of forests to uncontrollable fires.

What did it take for California to become a fire success story, rather than a failure?

Preventing California from descending into fiery chaos required permanent changes in individual people's lives, and permanent changes in the way Californians thought about their state. Changing how we build and where we build proved to be one of the biggest challenges of the 2030s.

"Building in a forest that has a history of wildfires probably [wasn't] the best idea," Lizzie Johnson, the fire reporter for the *San Francisco Chronicle*, told me years before California made its changes. "We need to stop assuming that humans can fight back nature, because at a certain point we can't. You have to acknowledge your own weaknesses."

According to Johnson, the key to this mind shift rested, as always, in the profound realization that we have the power to change. "Maybe it can be that moment in history books where they write, 'It almost got really bad, and once upon a time, California had these huge megafires.' Maybe people will look back on 2018 and study it, because there aren't any years that compare to it to study anymore."

Working together, residents of California and their local and state representatives implemented sweeping changes to how they tended the land and developed and preserved communities around it. While continuing its commitment to 100 percent clean energy and an economy-wide carbon neutrality by 2045, California's government also increased funding for statewide fire suppression and fuel-management programs. At the same time, residents in high-risk areas from Pasadena to Paradise took a more proactive approach to managing the land, reducing vegetation around their homes, and building with fire-resistant materials, such as locally sourced bricks and carbon-negative concrete. Such collective and individual actions helped better manage future fires and reduced carbon emissions, a consistent theme throughout the decade.

* * *

Because cities around the world continued to grow throughout the decade, we were able to use building and planning practices that simultaneously acknowledged the ongoing climate emergency and addressed its most-pressing implications. Continued efforts to reimagine urban spaces, from Atlanta to Seattle, concentrated on ripping up concrete from existing infrastructure, which proved to have immediate short- and long-term effects on reducing flooding and heat waves.

Cement—the primary ingredient in concrete—is the most widely used material in our civilization. It is made by heating up limestone rocks to super-high temperatures, which releases carbon dioxide twice over: as a chemical reaction by-product and as a result of burning fossil fuels to heat it.

Existing pavement in parking lots, roads, and buildings exacerbates the effects of extreme heat and intense flooding. Less concrete means more green space, which means more public gardens and fewer deaths from extreme weather. Denser housing, with green roofs, will make cities more walkable and car-free.

The quickest way to a carbon-free world, we realized, was to put an end to the practice of burning dead dinosaurs for heat and energy and to switch immediately over to electricity to power all forms of transportation, heating, and cooling. The more electricity we generated from carbon-free sources, the cheaper and more efficient the utility became, which helped produce better air and water quality and a wider distribution of energy ownership across the country. Such attempts presented new opportunities for labor and fostered the rise of new green industries, which revitalized previously marginalized communities with affordable and sustainable housing, and, increasingly, carbon-free infrastructure.

While emissions continued in hard-to-reduce industries, like iron and steel, a national commitment to decarbonizing these industries, which accounted for nearly 25 percent of all emissions in 2020, started to take shape during the 2030s. In the spirit of the age, we invested in alternative products to offset those emissions, despite the stubborn technological difficulties and growing expense. It was crucial to combat the carbon dioxide produced as part of the extraction and production of raw iron ore. We could no longer afford the luxury of inaction.

A centerpiece of the emerging circular economy, electric arc furnaces started to replace the outdated furnaces used in refineries, supplying a huge percentage of the world's demand for steel by recycling old steel as we always have, but now with 100 percent carbon-free electricity. A parallel boom in wooden skyscrapers using cross-laminated timber greatly reduced the demand for steel, quickly bringing the global economy into balance with what we can afford to produce while stabilizing the climate. Made up of small-diameter trees glued together, cross-laminated timber produces a strength similar to steel and concrete at a fraction of the weight. Using wood in this way actually turns buildings into places to store carbon—completely reversing the climate impact of the construction process. One study found that using wood to construct a 400 foot (120 meter) skyscraper could reduce the building's carbon footprint by 75 percent.

A spirit of experimentation and creativity characterized this decade of radical reinvention. Old paradigms and practices that maintained the status quo gave way to transformative practices that encouraged people everywhere to reimagine how the world could operate if we worked together. This spirit, which captured the hearts and minds of the public, was anchored in a collective sense of purpose in the face of an existential crisis, an international culture of empathy.

By the 2030s, we started to know that a better world was possible, because we were witnessing it coming to fruition all around us.

* * *

From democracy to the economy, we started to consider what a climate emergency means for our neighbors, no matter where they live. We made choices with their best interests in mind, not just our own, a selfless process that had seemed impossible just a decade ago. In hindsight, changing how we think about travel was one of the best things we did.

While phasing out all modes of gas- and oil-based transportation, we also started to phase out air travel entirely. Or at least air travel as we once thought of it. Airplane travel was the single most fossil-fuel intensive activity in the world, other than space travel. Globally, aviation used to be the fastest-rising source of emissions, and absent major changes, the aviation industry was set to grow another 700 percent by 2050—more than enough on its own to bust our efforts to keep the 1.5-degree climate goal. In the 2010s, a single round-trip cross-country flight emitted about as much carbon dioxide as the average person in India did in an entire year. The relative ease, efficiency, and affordability of air travel meant we were much more likely to choose that form of transportation. Planning a spontaneous cross-continental trip was often as easy as getting an Uber or ordering a meal on Seamless. And yet it was also a luxury increasingly hard to justify, even though most of our families and friends were spread all around the world.

The first step was banning first-class and business-class airline seats, which accounted for about 30 to 40 percent of all space on airplanes; that was followed soon thereafter by increased taxes on private jets. Eventually we realized that without lighter-weight batteries to offset the emissions of jet fuels, we needed to slowly implement a moratorium on air travel until a major breakthrough in battery technology emerged, but by that time, nearly every city in the world had convenient and affordable carbon-free transportation options to choose from. Even perceived solutions, like offsetting carbon by planting trees, proved problematic and unsustainable in the middle of the climate emergency. Occupying the same fields needed for subsistence agriculture in developing countries, these trees had to survive for decades just to ensure the carbon from a two-hour flight would stay out of the atmosphere. Quickly it became clear that from a justice perspective, the best option was to forgo air travel as a matter of solidarity, in the same way the Greatest Generation went without sugar during World War II.

Combined, these restrictions naturally led to a demand for fast, reliable, comfortable long-distance travel. Fortunately, the United States had the money to pay for it.

After a ten-year effort, investing hundreds of billions of dollars, the US neared completion at the start of this decade of a world-class, high-speed rail network and a burgeoning hyperloop system. Transatlantic passenger ships briefly made a comeback: the fastest ship, the *Greta Thunberg*, was able to make the trip from New York to London in just over two days. Traveling was fun again, an adventure in exploration and experience, if not efficiency, which we all enjoyed guilt-free and with pride. Meanwhile, immersive virtual-reality business meetings replaced face-to-face meetings, and even the technology started to compete with real-world vacations. Why spend the time and energy to move your body across an ocean when you could visit international cities like Paris, London, Tokyo—and interplanetary destinations like Mars—in an afternoon? But, of course, we continued to travel. When we do, the efficiency and automation of the circular economy, combined with a universal basic income and a four-day workweek, means almost everyone has time for slow travel now. The alternatives to air travel were so welcome we hardly miss flying. We relax on trains and ships, interact with people in a more friendly and much more spacious environment. The journey has started to trump the destination. In the future, we see the world and meet people everywhere, fostering connection and community, in ways we once thought impossible, or inconvenient. By reimagining the reasons and the means for long-distance travel, we revolutionized our transportation systems, considering—for the first time since the invention of the modern jetliner—how best we can move our bodies freely through space and time.

The world has become smaller, and we have grown closer in our communities. We exchange ideas freely and share spaces in novel ways that respect and foster a palpable sense of belonging and unity in a shared vision for the future Earth.

* * *

In 2035, global emissions finally started to sharply decline—down more than 50 percent from 2020 levels. Even though the temperature was still rising, we managed to avoid a 1.5-degree rise. We were in the middle of the Great Drawdown, a period of rebirth that allowed us to scale back emissions through individual and collective actions. Nowhere were these efforts more apparent—or more necessary—than our approach to agriculture.

Far beyond driving and flying, beyond heating and lighting our homes, even beyond steel and cement production, the leading impact of humanity's presence on this planet is from what we eat. Intimately connected, soil, water, food, and carbon form the basis for our ten-thousand-year-old human experiment with cultivating plants. Community-controlled, biodiversity-based food systems used to be the norm around the world, until the exploitation of fossil fuels, which kicked off during the twentieth century an arms race of mechanization and increasingly corporate agriculture. As soon as land was viewed as investment income for hedge funds, it was managed to maximize quarterly profits, not the health of the soil or the quality of the foods grown on it. Despite amazing technological developments during the first half of the twenty-first century—fields in Arizona are laser-leveled using GPS to prevent even a drop of wasted water—commodity agriculture was so unimaginative for so long, relying on the same crops grown in basically the same monoculture fashion, which ravaged local ecosystems and wreaked havoc on the world's environments.

Agriculture, particularly animal agriculture, has utterly remade the Earth's surface. It remains one of the worst culprits in bringing us to the brink of a mass extinction, consuming half of the world's arable land and using 80 percent of the world's fresh water supply. At the start of the 2020s, as India struggled to address its depleted levels of groundwater, the number of Indians living without regular access to potable water approached half a billion. Meanwhile, large-scale floods, droughts, and heat waves threatened crop yields around the world, and as the rate of evaporation increased exponentially the hotter the temperature got, farmers were using even more of the world's water supply just to produce the same amount of crops. Warming temperatures in the United States spiked the demand for water to unsustainable levels, and as the weather became more erratic, farming communities used to predictable levels of precipitation were forced to tap into groundwater in search of something resembling agricultural stability. The uncertainties in agriculture pushed thousands of farmers to suicide in both the US and India.

Because of human activity, the planet's soil, water, carbon, and nutrient cycles were almost unrecognizable by the start of the twenty-first century. Biogeochemistry, the science of these cycles, began to point toward the early stages of a complete breakdown of the planet's life-support systems on par with a mass extinction. Comparing arable conditions to the state of the oceans 252 million years ago, biogeochemists warned of another Great Dying: a dire warning the world finally started taking seriously.

From America's heartland to the fields of Uganda, we started to learn how to survive under planetary constraints by practicing a form of agriculture that was regenerative, not extractive.

A rapid transition away from industrial animal agriculture, including a boom in the quality of plant-based meat alternatives, set the stage for a rethink of farming. Just as fire is essential for the health of woodland ecosystems, grazing animals are necessary for the health of grasslands. But the way we'd been going about it for the past one hundred years was all wrong.

In Kansas, the seasons are named after the farming activities we perform—haying season or calving season or harvesting season—holdover terms from a time when we worked with nature instead of trying to manipulate its cycles. While there was always a space for subsistence hunting and small-scale rearing of large animals, farmers started throwing away the failed models of the past to produce a better future. Conversations between trusted friends about the weather occurred out in fields and in pickup trucks, words of advice and encouragement among people who shared a history together. These familiar conversations were necessary in helping farmers transition from industrialized agriculture to a more permanent, perennial relationship with the land, which also ushered in the animal husbandry and plant ecology of a new era of life on Earth.

Around the world, growing peasants-rights movements sought to wrest control of the land from corporations and return it to the people who lived and worked there. Smallholder subsistence farmers—in numbers totaling two billion strong—reasserted their rights to their own lands and started to practice agricultural systems that worked best for them and their communities.

Similarly, researchers in Kansas aimed for a completely different vision for agriculture. The Land Institute, based in Salina, had been working for decades on the perennialization of farming, moving toward ecological and social systems that would magnify the efforts of the people, plants, and animals. Thinking of farming less as an industrial production and more as care work, the institute bred alternative, perennial varieties of commodity grains like wheat and corn, in an attempt to imagine how they'd fit within future Kansas communities that worked alongside nature in a more holistic way.

Across the High Plains in the late 2030s, native grasslands were restored, and because grasslands more quickly store carbon and for longer periods of time than forests, landowners started earning income by storing carbon in their fields' deep root systems. An economic boom followed. Farmers invested their surplus in wind and solar farms. From Kansas to Canada, rural towns that long ago had been declining started springing back to life.

We were then well on our way to implementing the expanded Green New Deal for rural America—a reimagining of what healthy ecosystems and healthy communities look like when they're working together. Farming is a commitment to the future, after all. It's a tangible way of sending gifts to your descendants. Even non-farmers participated in this commitment. Around the country, people started transforming lawns in cities as diverse as El Paso and Sherman Oaks. Back in 2020, lawn grasses were the country's number one irrigated crop, more than corn, wheat, and fruit orchards combined. Lawns contributed tens of millions of pounds of fertilizer and pesticides to the environment each year and used roughly half of all urban water in the US. By repurposing them, we improved at once the availability and quality of our water supply, carbon storage, food production, and our personal enjoyment of our surroundings. Actions as simple as planting a chestnut tree in a yard provided twenty pounds of food per year, a ten-thousand-dollar value over fifty years. And a lawn planted in creeping thyme produced a walkable, beautiful landscape that required no water to maintain.

On a larger scale, America's newest national park, a thousand-mile-wide stretch of plains from Texas to the Dakotas, populated with free-roaming herds of buffalo, now stored an additional hundred million tons of carbon per year and protected the dwindling Ogallala Aquifer for future generations.

The era of regenerative agriculture that followed our urban reimagining mirrored the cooperative and mutually beneficial structure of local ecosystems and ushered in the kind of generational work that helped reverse the effects of climate change, here and abroad, in big and small ways alike.

Back in 2018, Aubrey Streit Krug, director of the Land Institute's Ecosphere Studies project, described to me what an effort to perennialize agriculture might look like: "We would see changes in land use and land cover. And we would also see, I think, people in collective movements . . . living and learning in diverse communities that have the ability to persist season after season. Just like the prairie. . . . I think if a turn toward perennialization can happen anywhere, why not here?"

## 2039: LIFE WITHOUT ICE

Nothing, however, compared to the continued deterioration of the planet.

Throughout the 2020s and 2030s, as we worked to rebuild society in a more equitable and just manner, we were still locked into the level of climate change we had caused because of our past recklessness and extractive practices. Anticipating what could still go wrong remained a centerpiece of our commitment as we continued to come up with resiliency plans that left space for the unthinkable, including the collapse of the major ice sheets in the Arctic and Antarctic.

Warming in Alaska, along with rising temperatures throughout the Arctic, accelerated as the loss of snow and ice cover set off a feedback loop of further warming. Since the turn of the century, Arctic temperatures in the winter rose more than 4 degrees Celsius (7.2 degrees F). Globally, temperatures briefly crossed the 1.5-degree Celsius mark, even though human emissions had drastically declined. The lag time in the climate system, combined with additional emissions from melting permafrost in the Arctic, had finally caught up with us. During the summer of 2039, there was no ice in the Arctic Ocean for the first time in more than a hundred thousand years.

Along North America's Arctic coast in Utqiaġvik (formerly Barrow), Alaska, the change to a new climate had come nearly overnight. Thunderstorms occurred regularly in the tundra. During 2016, the warmest year in Utqiaġvik's recorded history, the city endured an incredible stretch of one hundred consecutive days of above-normal temperatures. As the sea ice retreated from shore, humidity from the open ocean flooded in, and the tundra began to spontaneously support a whole new ecosystem of plants and animals.

Throughout the Arctic, warmer oceans helped melt sea ice, and this heat influx shifted the jet stream. Weather throughout the entire Northern Hemisphere permanently changed in response. Evidence showed that Hurricane Sandy, which brought an unprecedented flood to New York City, could only have happened the way it did because of climate change in the Arctic.

As warm weather patterns became stuck over the planet's polar regions, key glaciers in Antarctica and Greenland experienced an accelerated rate of melting. Rather than the century-long projection scientists had anticipated, these glaciers were expected to melt within decades. Rapid freshwater melt from Greenland and Antarctica threw a wrench into the main oceanic conveyor belt circulating heat between the tropics and the temperate midlatitudes, which created a surge of superstorms like Sandy as the atmosphere attempted to compensate.

In a remote region of Antarctica known as Pine Island Bay,

2,500 miles from the tip of South America, two glaciers—Pine Island and Thwaites—continued to hold human civilization hostage. Stretching across a frozen plain more than 150 miles long, Pine Island and Thwaites are two of the largest and fastest-melting in Antarctica and have for millennia marched steadily toward the Amundsen Sea, part of the vast Southern Ocean. Farther inland, the glaciers widen into a two-mile-thick reserve of ice covering an area the size of Texas. Together, they act as a plug, holding back enough ice to pour eleven feet of sea-level rise into the world's oceans—an amount that would submerge every coastal city on the planet.

Finding out how fast these glaciers were collapsing proved to be one of the most important scientific questions facing the world. Scientists the world over prepared for the most likely worst-case scenarios. They looked back to the end of the last ice age, about eleven thousand years ago, when global temperatures last rapidly warmed. The bad news was, scientists found growing evidence that the Pine Island Bay glaciers had collapsed rapidly back then, flooding the world's coastlines—partially the result of something called "marine ice-cliff instability." Because the ocean floor is deeper beneath these glaciers in this part of West Antarctica, each new iceberg that breaks away exposes taller and taller cliffs. Ice gets so heavy that these taller cliffs can't support their own weight, making destruction inevitable.

In 2020, every marine glacier that rests on solid rock still had a floating shelf of ice attached. An ice shelf is one of the best diagnostic tools we have to measure the health of the parent glacier. If the rate at which an ice shelf fractures and forms icebergs—a normal part of every marine glacier's life—suddenly speeds up, the glacier is most likely destabilizing. Between 2005 and 2015, fracturing of the Thwaites ice shelf sped up by a factor of three—a worrying sign for sure.

In 2039, the Thwaites ice shelf began to collapse entirely, directly exposing the glacier to the wave energy of the Southern Ocean, the wildest body of water on Earth. The warm water current that encircles Antarctica made a detour into the void left by that ice shelf, and a collapse of Thwaites Glacier itself proved imminent.

Scientists were worried that the most frightening collapse scenario, which may have happened at Thwaites at the end of the last ice age, could kick in if warm water eroded the ocean-facing cliff of the glacier to such an extent that skyscraper-size shards of ice began to fragment off nearly continuously. Ice isn't the strongest of building materials, and structural engineers that treat glaciers like giant buildings have found that for heights greater than three hundred feet, a sheer ice wall will inevitably collapse—no matter its temperature. At its heart, Thwaites is more than six thousand feet thick. Once it started to collapse in earnest, Thwaites retreated dozens of miles in just a decade or so—a geological blink of an eye. Gradually, a pile of massive icebergs at Thwaites's new outlet formed a temporary dam to slow its continued demise. There was simply no room for all the ice to spread out into the sea at the rate the glacier was breaking apart.

The in-person spectacle of all of this approached the limits of human comprehension. On present-day boat tours in Alaska, the calving fronts of marine glaciers are typically no more than a few hundred feet wide. In Greenland, Jakobshavn Glacier, the fastest-retreating glacier in the world during the 2000s, is about two miles wide. Jakobshavn's cliff face is about three hundred feet tall—exactly the height scientists predict is the maximum limit of ice cliffs on Earth. Occasionally, icebergs calving off Jakobshavn register on nearby seismographs—the size and weight of an entire city block's worth of apartment buildings: millions of tons. In 2040, the calving front at Thwaites is a hundred miles wide, from horizon to horizon. The only people able to see it are flying overhead in the International Space Station.

On the shorelines of New York City, London, Mumbai, Manila, and countless other coastal cities around the world, people experienced the partial collapse of Thwaites through a series of imperceptibly higher tides, one after another, for several years. High tide arrived and departed like normal, but a bit of that water just didn't leave. Billions of tons of ice, adrift in massive floes spreading out from West Antarctica, drifted northward toward warmer waters, melting again after a wait of millennia.

Sea walls, if they have been constructed, were snapped closed. Where coastal barriers didn't already exist, it was too late to build them with much foresight or intentionality—the political fights over whose homes and businesses were on the "wrong" side of the levees were contentious. Some places, like South Florida, where the porous bedrock geology didn't allow for sea walls, were partially abandoned. In the Netherlands, a country that has been holding back the oceans for centuries, this proved too much. After a Thwaites collapse, water from the Ems, Meuse, Rhine, and Scheldt Rivers needed to be pumped uphill to reach the ocean. In the Marshall Islands, the waves began to erode the shorelines with renewed intensity. Around the world, 100 million people had to find new homes.

Thwaites's partial collapse surged global sea levels and helped to destabilize other marine glaciers too—ironically in Greenland—sending giant icebergs streaming away from Antarctica like a parade of frozen soldiers to inflict a global catastrophe the likes of which we've never seen. As West Antarctica lost the weight of so much ice, local sea levels there actually dropped because the pull of gravity was less. That water was redistributed around the world's oceans and shorelines.

Instead of a three-foot increase in ocean levels by the end of the century, six feet was now more likely, in line with predictions made by climatologists Rob DeConto, at the University of Massachusetts Amherst, and David Pollard, at Penn State University, back in the 2010s.

DeConto and Pollard's breakthrough came from trying to match observations of ancient sea levels at shorelines around the world with current ice sheet behavior. Around three million years ago, when global temperatures were about as warm as they're expected to be later this century, oceans were dozens of feet higher than today. Previous models suggested that it would take hundreds or thousands of years for a sea-level rise of that magnitude to occur. But once they accounted for marine ice-cliff instability, DeConto and Pollard's model pointed toward a catastrophe if the world maintained a business-as-usual approach or if feedback loops kicked in earlier than expected.

Next to a meteor strike, rapid sea-level rise from collapsing ice cliffs is one of the quickest ways our world can remake itself. This is about as fast as climate change gets.

"Every revision to our understanding has said that ice sheets can change faster than we thought," Jeremy Bassis, an ice sheet scientist at the University of Michigan, told me. "We didn't predict that Pine Island was going to retreat, we didn't predict that Larsen B was going to disintegrate [in a matter of months in 2002]. We tend to look at these things after they've happened."

That was a recurring theme throughout these scientists' findings in Antarctica: what we do today will determine how much further Pine Island and Thwaites collapse in the future. We can still prevent this dire scenario before it's too late.

* * *

By 2040, the world cut its emissions by two-thirds of current levels, but even with that reduction, temperatures continued to rise. Even a partial collapse of the glaciers in Antarctica and Greenland created a sea-level emergency. Leaders of the world started to debate ways to stabilize temperatures and extreme weather. They considered everything, including the unthinkable: geoengineering.

# 2040–2050: NEW TECHNOLOGIES AND NEW SPIRITUALITIES

*All that you touch you Change.*
*All that you Change changes you.*
*The only lasting truth is Change.*

—Octavia E. Butler, *Parable of the Sower*

A trope of sci-fi movies these days, from *Snowpiercer* to *Geostorm*, is that our failure to tackle climate change will eventually force us to deploy an arsenal of unproven technologies to cool the planet. Think sun-deflecting space mirrors or chemically altered clouds. And because these are sci-fi movies, it's assumed that these grand experiments in geoengineering will go horribly wrong.

But as emerging studies have continued to suggest, the fiction is much closer to reality than we previously thought.

When most people hear "climate change," they think of greenhouse gases overheating the planet. But there's another product of industry changing the climate that has received scant public attention: aerosols. Aerosols are microscopic particles of pollution that, on balance, reflect sunlight back into space and help cool the planet down, providing a crucial counterweight to greenhouse-powered global warming.

An effort to co-opt this natural cooling ability of aerosols has long been considered a potential last-ditch effort to slow down global warming. The promise of planet-cooling aerosol technology has also been touted by techno-optimists, Silicon Valley types, and politicians who aren't keen on the government doing anything to curb emissions. "Geoengineering holds forth the promise of addressing global warming concerns for just a few billion dollars a year," wrote Newt Gingrich in an attack on proposed cap-and-trade legislation back in 2008.

But there's a catch.

This surplus of aerosols is a huge problem. At high concentrations, these tiny particles are one of the deadliest substances in existence, burrowing deep into our bodies, where they can damage hearts and lungs.

Air pollution from burning coal, driving cars, and controlled fires to clear land, as well as from other human-related activities, is the fourth-leading cause of death worldwide. It kills about 5.5 million people each year. Nearly everybody is at risk, with roughly 92 percent of us living in places with dangerously polluted air. That alone makes reducing air pollution a necessary goal.

Natural aerosols—bits of dust, salt, smoke, and organic compounds emitted from plants—are an integral part of our planet's atmosphere. Without these types of aerosols, clouds would likely be unable to make rain. But, as is the case with greenhouse gases, human activity has produced too many aerosols in the form of excess air pollution. The bulk of the human-emitted aerosols linger in the lower atmosphere, which worsens their impact on our health. The result is a devil's bargain: we need aerosols for normal weather and to help moderate rising temperatures, but they are also killing us.

We might be locked in this deadly embrace for longer than we'd like. The cooling effect of aerosols is so large that it has masked as much as half of the warming effect from greenhouse gases. There's no way around it: aerosols can't be wiped out without dramatic consequences. Take them away and temperatures would soar almost overnight. In the 2030s, after nearly two decades of radical emissions reductions around the world and the cleansing of our skies of air pollution, the effect of the lost aerosols continued to drive global temperatures upward, perhaps to dangerous levels.

People have been aware of the influence of aerosols for centuries. In the 1200s, Londoners complained about the clouds of coal smoke. In 1783, Benjamin Franklin observed that tiny particles from volcanic eruptions tended to chill the weather. Throughout the late 1800s and early 1900s, dense smoke from coal blocked out daylight in Chicago, Pittsburgh, Saint Louis, and scores of other cities. In 1991, Mount Pinatubo in the Philippines erupted, providing a natural laboratory for studying aerosols' impact. The resulting research gave scientists solid evidence that particles in the atmosphere tended to cool the planet, essentially proving Benjamin Franklin's hunch two centuries earlier. During the first part of the twenty-first century, scientists continued to puzzle over exactly how aerosols from tailpipes and smokestacks alter the weather, in part because the particles are incredibly difficult to study. To research them, scientists sought out remote corners of the globe far from industrial pollution, like the seas around Antarctica.

Aerosols are much bigger than air molecules, so they weigh more and tend to fall out of the sky within days or weeks after they're released. There's also a ten-thousand-fold range in their sizes and a wide variety of sources, making their behavior relatively unpredictable. Black carbon aerosols from forest fires, for example, tend to suppress cloud formation by warming the air, which makes tiny water droplets evaporate. Similarly, sulfate aerosols from burning coal can make clouds grow bigger and rainstorms stronger. Thunderstorms in China vary on a weekly cycle, right in line with local factory schedules.

What's clear is that, on balance, aerosols are cooling us off. If we magically transformed the global economy overnight, and air pollution fell to near zero, we would experience an immediate rise in global temperatures of between 0.5 and 1.1 degrees Celsius. (For reference: as of 2020, the climate has warmed about 1.2 degrees Celsius since the start of the Industrial Revolution in the nineteenth century.) The warming would be concentrated over the major cities of the Northern Hemisphere, close to where most aerosols are emitted. In the hardest hit parts of highly urbanized East Asia, for example, the complete removal of aerosols would likely have a bigger effect than all other sources of climate change combined. Temperatures in the Arctic could jump as much as 4 degrees Celsius (7.2 degrees F)—a catastrophe that would shove the region further toward a permanently ice-free state. Research in 2019 showed that this effect might last only five or ten years, but that might be enough to push already fractured glaciers beyond a tipping point, with disastrous consequences.

So what do we do?

* * *

Previous attempts at removing harmful aerosols have proved largely successful, especially in the United States and Europe. The US Clean Air Act, one of the most important fruits of the 1970s environmental movement, led to a sharp and nearly immediate fall in air pollution, likely saving millions of lives. "This is known territory, at least compared to massively reducing CO2 emissions," Bjørn Samset, research director at Norway's Center for International Climate Research, told me. Pumping artificial aerosols into the upper atmosphere should also work, in theory. Balloons and airplanes could spray benign aerosols like calcium carbonate (essentially crushed limestone), which would be carried throughout the upper atmosphere by the wind. One recent study estimated it would take 6,700 business jet flights per day—outfitted with spraying equipment—to keep enough aerosols in the stratosphere to cool the climate by 1 degree Celsius (1.8 degrees F). The cost: $20 billion per year, more or less in line with Gingrich's estimate from a decade ago, adjusting for inflation.

But nothing of this scale has ever been tried. In fact, the evidence is that messing with aerosols has already led to past periods of rapid warming. After the Clean Air Act was passed, global temperatures began climbing in the late 1970s, ending a relatively stable thirty-year period. A similar pattern has now begun in Asia. In recent decades, the rapid economic rise of coal-powered China and India, coupled with the resulting aerosol emissions, has blackened skies in Shanghai, Delhi, and other megacities. This almost certainly has contributed to a slowdown in the rate of warming, globally. In the 2010s, China responded to public outrage over the country's air-pocalypse by putting in place pollution controls. And there's initial evidence that they're beginning to work. India, meanwhile, has taken the dubious title of having the worst air quality in the world, and outrage is starting to grow there too.

Bjørn Samset thinks the immediate health benefits of curbing air pollution mean that China will likely stick to these efforts, in spite of the potential warming effects. "It's very plausible that Asian aerosol cleanup—which saves lives directly by reducing air pollution—can get prioritized over strong greenhouse gas cuts," he explained. What was once the realm of scary science fiction and conspiracy theory is now entering the mainstream of atmospheric study—only those now conducting the experiments are clear about the risks.

"Geoengineering is like taking painkillers," said Frank Keutsch, a Harvard chemist who's working on the problem. "When things are really bad, painkillers can help but they don't address the cause of a disease and they may [do] more harm than good. We really don't know the effects of geoengineering, but that is why we're doing this research."

And even if geoengineering with aerosols works to offset warming? That, too, could have disastrous side effects.

Samset told me that embarking on a planetary-scale aerosol geoengineering project would produce "a wide range of unintended regional consequences." One of the biggest risks, according to a study published in *Nature Ecology and Evolution*, is that the cooling would work too well, producing shifts in ecosystems at "unprecedented speeds"—the kind of scenario that was dramatized in the movie *Snowpiercer*. That could be a fatal shock to animals and plants already stressed by decades of warming.

"I could imagine global conflicts breaking out over these types of actions," Susanne Bauer of the NASA Goddard Institute for Space Studies, told me. "On the other hand, I do believe geoengineering must be studied, just to be aware and educated about the possibilities."

## FROM CRISIS TO CHANGE

The thirty years from 2020 to 2050 will be among the most transformative decades in all of human history. Collapsing ice sheets, the aerosol crisis, and rising sea levels will force more people to leave their homes than at any other point in human history. In some places, that means conflict is inevitable.

A study from researchers at the University of California at Berkeley found that higher temperatures and shifting patterns of extreme weather can cause a rise in all types of violence, from domestic abuse to civil wars. In extreme cases, it could cause countries to cease functioning and collapse altogether.

This ominous reality of climate change is far from fated, however. A rapidly changing environment just makes conflict more likely, not inevitable. People, ultimately, are still in control. Our choices determine whether or not these conflicts will happen. In a world where we've rapidly decided to embark on constructing an ecological society, we'll have developed countless tools of conflict avoidance as part of our climate change adaptation strategies.

Still, there will be those who choose to live outside the mainstream society who may pose an existential threat to the rest of us. Some groups and a few rogue countries will try to prevent the rest of the world's transition toward ecological and social justice. They will do this either because of the lingering influence from the dwindling fossil fuel industry, or because of a fascist ideological response to climate change that puts human rights at risk, or out of desperation.

Mary Annaïse Heglar, a climate essayist and advocate for intersectional approaches to racial and environmental justice, is inspired particularly by Octavia Butler's *Parable of the Sower* for an example of how things could go very badly. In the book, Butler describes fire-obsessed cults that spring up in a post-rapid-climate-change world, craving some sense amid the destruction and chaos they see all around them. Heglar thinks that could be just the beginning.

"The future I see is really ugly unless something very, very drastic changes," Heglar told me. "It's a world where people find many, many different ways, very creative ways, to be cruel to one another. Unpredictability brings out people's cruelty if you're not careful. And most people are not careful."

Heglar specifically thinks of the racial massacre in East Saint Louis, Illinois, in 1917 as an example of the kind of violence that might emerge if the world is not careful. Angry white mobs murdered dozens of Black people after they were hired in place of striking workers at factories during World War I. If lifesaving technology is not distributed fairly, or if governments lean too heavily on austerity along racial lines, or if climate disasters fragment already vulnerable populations, the result could be truly ugly.

"So many things that we think are impossible today could be completely normal in twenty years," Heglar told me. "I hear people saying now that 'when it gets really bad, I'll just move to New Zealand or I'll move to Sweden where climate change impact is not going to be that drastic.' But it's not going to be cute there. First of all, it's going to be mostly the one percent living there. So if you think your regular ass is gonna be able to buy land in New Zealand, good luck."

An escapist attitude is probably the most dangerous reaction to climate change today. It drives to the heart of how the problem of climate change came into being in the first place: by imagining ourselves as individuals who somehow exist outside the context of an interconnected, living ecosystem on a planet where all of our actions deeply affect one another, we fail to see each other's humanity and right to simply exist. It's the same attitude that drives the richest men in the world today to create their own private space agencies. Those who are already being affected by the climate emergency can't and won't simply be left to fend for themselves while the privileged few plot their escape plans—to higher ground in their neighborhood, to inland mountain refuges, to Mars.

Until we build a world that works for everyone, we'll continue to have people whose survival is systematically erased by those in power. That's the dystopia for the rich and powerful: a world where the rest of us finally realize the power we had all along to fight for a justice-focused society.

It will take active, conscious effort to defuse the tensions sure to arise in a warming world. Overcoming a coordinated effort by the fossil fuel industry to save itself is not going to be easy, but we know it's coming. That effort has been going on since the fossil fuel industry began, and it won't just go away in the 2040s, even amid two decades of radical and hopeful changes. As always, our best hope will remain that we can prepare along the way to increase the chances of a peaceful transition to a fossil-free world.

We know that the weather in the 2040s will be worse than it is today. A major, sudden change, like a collapsing ice sheet or a quick rise in global temperatures after eliminating aerosols, would make the weather even more destructive than current predictions, even if we are able to radically reduce greenhouse gas emissions. What we can control, of course, is how we decide to respond to the worsening weather.

Since my conversation years ago with Rear Admiral Titley, I've repeated his idea of "catastrophic success" over and over to myself when I think things can't get any worse, and I've let it shape my view of how the world could quickly change beyond our wildest imaginations—for the better. Titley sees the warming world both as a scientist (he's a meteorologist, by training) and as a former military officer. He understands that the potential for a massive increase in refugees is a heartbreaking and almost inevitable looming humanitarian crisis due to the science of the escalating severity of droughts, floods, and severe weather we've already seen in recent decades and the historical tendency for leaders to close borders during times of crisis. A worsening of this trend could make the world practically ungovernable in our lifetimes.

The US military has been among the first large-scale entities to recognize this. That kind of makes sense if you consider their mission of ensuring US safety and prosperity continues for as long as possible: without planetary stability, there is no US stability. That's part of why US military strategists at the Pentagon have begun calling climate change a "threat multiplier."

When Titley talks about migration, though, even he struggles to put the stakes in context. In the 2040s, if global sea levels rise by three feet and droughts, fires, heat waves, and floods continue to worsen, we could see around 250 million people forced from their homes. That's about four times as many people as are currently displaced, and about fifty times as many as were displaced during the Syrian Civil War. In short, it would challenge our understanding of nationality, borders, and politics as usual.

"Post–World War II," Titley told me, "tens of millions of people within Europe were on forced migration in the 1940s. We kind of gloss over that part of history. I mean, Europe was really bad after World War II. It's part of what got the Marshall Plan. I think it really kind of scared us that, hey, this whole place is just collapsing, basically, and something had to be done."

An uncontrolled, unanticipated climate-related migration crisis could be even worse than the refugee crisis after World War II, which, despite its horrors, displaced less than one percent of the world's population. Climate change could displace three times that amount just in the next two or three decades. Although displacement due to extreme weather is already becoming increasingly common, the proximate cause of displacement and migration is usually fleeing violent conflict. How do we anticipate a world that could quickly fracture, and urgently work to reduce the risk of violent conflict before it occurs?

A crisis like this will require proactive harm reduction on a civilizational scale. We will need to establish policies that encourage, rather than restrict, freedom of movement. And we must establish robust social safety nets so that families are less likely to abandon their homes in search of a place where they can simply live. Also, even before we reach zero emissions globally, we will have to recognize the need to take aggressive actions to reduce the level of carbon dioxide in the atmosphere. All of this will remain just as urgent in the 2040s as in 2020.

"I'm probably wrong," said Titley, "but I'm actually more optimistic that we are going to do real things now than I have been for a long, long time. I think there's actual legitimate cause for optimism."

Specifically, Titley pointed to the steady shift away from outright denial among rank-and-file members of the Republican

Party as evidence that attitudes can shift toward action, no matter how meager. And once that facade of climate denial breaks, an avalanche of action could soon follow. "We may be much closer to catastrophic success right now. Things can change, and not always for the worse. They can change for the better. It can happen very, very quickly."

## 2040–2042: SHAPING THE FUTURE

By the 2040s, we achieved a carbon-free society in the United States, Europe, and many other places throughout the world. We began to draw carbon back out of the atmosphere in huge quantities in the oceans, soils, grasslands, and forests. We turned the corner toward an ecological society, because we prioritized justice and the inherent dignity of every living being on the planet. A revolution in our mindset and our relationship with the Earth allowed us to recognize that we all have value—that everyone and every species deserves the right to exist.

After a three-decade struggle, we realized that we deserve a beautiful, pleasurable, and justice-centered world, and we started to liberate one another and ourselves. To paraphrase author and women's rights activist adrienne maree brown, we shaped the future we longed for and had not yet experienced.

We did this because our world was dying. The thirty years between 2020 and 2050 was a period of shared grief and loss for so many of us, but it inspired us to meet these hardships with courage and hope and imagination, because catastrophic setbacks became the norm. Together we started the hard work of building a new world and continued to do it because we had to—for Indigenous folks after centuries of erasure and oppression, for Pacific Islanders fighting to protect their islands from sinking forever beneath the waves, for the very basic idea that there was no future worth fighting for that was not rooted in justice.

Farhana Sultana, a political ecologist whose work focuses on water rights, told me that despite her native Bangladesh being written off as a catastrophe in the making during the 2010s, the mood in her home country remained "stubbornly optimistic" because "we've had no other choice but to be so to survive."

*Stubbornly optimistic*. That pervasive attitude allowed us to endure the worst effects of climate change and continued well into the 2040s, when technology and innovation started to open doors that were previously shut. How we deployed these technologies—and determined who benefited from them—decided our fate as a civilization.

A techno-fix climate future would be as equally oppressive as today's capitalist utopia if it weren't coupled with radical decolonization and proactive efforts to extend climate reparations to the billions of people who endured the harshest impacts of climate change. The mechanisms by which we're able to ensure justice proved to be the most important "technological" revolutions of this century.

This—and our collective work over the past decades—was put to the test, after the Thwaites glacier partially collapsed in 2039. Its sudden loss caused sea levels to rise three feet in the span of a decade, faster than most scientists thought possible. Because of a quirk in the Earth's gravity, sea levels rose highest on the East Coast, permanently putting parts of Miami, Charleston, Norfolk, Philadelphia, New York, and Boston underwater.

Since the 2020s, when the world finally took collective action against the looming crisis of climate change, we consistently and correctly chose the path of ecology and justice. The flooded cities were as prepared as they could have been. In the case of Miami, that meant a combination of adaptation and retreat. In time, life in Miami started to resemble life in other archipelagos, an interconnected and flourishing community that embraces its watery reality, like the Florida Keys.

At heart, the issue of climate change is about survival. Specifically, as this century unfolds, it's about who gets to survive and who doesn't. During these decades of radical change, every decision needed to be made as if it were life or death. Because it always was. That kind of wild, constant, existential, emotional labor would have been impossible for most people to manage if we hadn't already spent decades actively working to develop new ways of caring for one another.

World leaders signed a global climate migration treaty in 2040, allotting proportional residence of refugees according to the historical emissions footprint of every country on Earth. By establishing a permanent visa program for the 100 million people directly affected by climate disasters, the treaty effectively abolished national borders. Nearly one-quarter of these refugees found new homes in the United States.

In the early 2040s, a simultaneous breakthrough in desalinization technology and diplomacy in the Middle East set off a chain reaction of effort to defuse the migration crisis that had gripped the world since the partial collapse of the glaciers in Antarctica and the sudden surge in warming after we phased out aerosols.

We understood that capitalist consumerism propped up the oil-rich economies, and once they went away, the rest of the world agreed to support the former oil states in their transition to a circular economy. The former oil states were already well on their way to becoming leaders in manufacturing solar equipment, high-speed trains, and desalinization designs, but during an emergency global summit in Dubai, world leaders agreed on a plan to enhance cooperation on an aggressive carbon drawdown and began discussions to geoengineer the climate, a temporary endeavor that would be phased out gradually as carbon dioxide levels were drawn back below 350 parts per million. This approach proved consistent with a long-term stabilization of climate at levels that would no longer risk a large-scale breakdown in society.

This global effort for disaster preparedness and prevention extended the gains of a newly resilient circular economy in Europe and North America to the entire world. Though such efforts should have happened sooner, even at this late hour, the world came to an agreement: no matter who is negatively impacted by the effects of geoengineering, all countries acting in unison will ensure fairness. In one voice, international leaders declared that meeting everyone's basic needs is nonnegotiable. Together we reduced the reasons for conflict and at last built a society where everyone on Earth could flourish.

Though global sea levels continued to rise, it still remained within our power to prevent a wholesale collapse of the massive ice sheets in Greenland and Antarctica. With careful planning and dialogue between communities most likely to be affected by climate change, the world's governments were able to avoid further economic collapse. Between 2020 and 2050, we learned how to work with one another and for one another. We matured our democratic systems of governance to ensure our ability to make difficult decisions quickly. And we grew comfortable embracing change, because we knew, each and every one of us, that we were building a better world.

Despite these challenges, the 2040s was a decade of regrowth—not of the extractive fossil fuel economy, of course, but of the plants, animals, ecosystems, and communities of people that have been stunted for so long by the status quo.

Once some form of geoengineering became inevitable, we established clear rules on who benefits from any geoengineering program. What would be the program's measures of success? If coastal property values were propped up because the ice sheets in Greenland and Antarctica slowed their collapse, would that boost to the global economy be fairly distributed to the farmers in Senegal or Paraguay whose crops would be destroyed because of an extreme weather event that likely wouldn't have happened without geoengineering? How would we measure these effects—and hold rogue profiteers to account?

At the same time, the slow lag of ocean heat storage, locked in from years of poor decisions, continued to escalate extreme weather events well into the 2040s, which threatened our radical transformation of society and put at dire risk millions of people living in small island states, in low-income neighborhoods, along riverbanks, and in farming regions around the world. To preserve the homelands of these millions, we ramped up our geoengineering capabilities, deciding, after another global-scale nonviolent wave of protest, that the patents to all carbon-free energy technologies should be made freely available to anyone.

* * *

Reducing emissions to zero is the best way to slow down climate change. We understood this in 2020 just as well as we did in 2045. The outstanding question was: What if after already greatly reducing our global emissions, the climate tipping points we previously set in motion are triggered anyway?

Technology is just the practical application of scientific knowledge—and that knowledge tells us that truly transformative solutions are the most practical way forward during an emergency like this. Because the techno-fixes that capitalism wanted us to devote public money to failed to keep our planet safe, few believed such an approach would work in this decade either, as the effects of climate change continued to inflict injustices around the world. Sure, we continued to take full advantage of technological improvements to produce food, energy, water, and shelter, but we treated those technologies with caution, finally understanding that celebrating technology as a goal in and of itself was dangerously misguided.

A revolution in all aspects of society offered us the best chance to build a livable world in our lifetimes—far more effective than any particular piece of renewable energy technology or tree-planting initiative.

If the rapid changes of the 2020s and 2030s taught us anything, it was that social movements are the best "technology" we have to bring about rapid and far-reaching decarbonization in all aspects of society. By bringing a fairer, more justice-centered world into being and an economy that fundamentally prioritizes planetary health and equality, we achieved a "technological" breakthrough that no amount of twentieth-century research and development funding ever could have imagined.

By midcentury, the world was a vastly different place. Some places became uninhabitable, because drought, flooding, and intense heat created wild, inclement weather. Elsewhere, entire nations became carbon-free, and a wholesale transformation was under way to convert as much as one-third of the planet's arable land into trees and other perennial plants that could be grown for food and fuel and draw carbon out of the atmosphere as quickly as possible. But we still needed to resort to other, even bolder methods to stabilize the climate.

Even after the global revolution of the 2020s and 2030s put our society on a path toward a completely decarbonized and circular economy, we were faced with a lot of difficult choices. Creating a truly sustainable world required ingenuity, creativity, and patience as we embarked on a completely different approach for the living planet we were stewarding.

## 2043–2045: NEGATIVE EMISSIONS

By the 2040s, we not only ran an entirely carbon-free electric grid but also scaled up technologies designed to actually suck carbon dioxide out of the atmosphere on an enormous scale. By 2045, carbon dioxide levels were approaching 500 parts per million, but we were on our way toward a return to the safe zone of planetary stability.

Our need to draw down carbon from the atmosphere was an expensive, but necessary, consequence of our delay in reducing emissions. Each part per million of carbon dioxide in the atmosphere amounts to about 2 billion tons of carbon. Extracting more than 100 parts per million from the sky—about as much as we've currently overshot safe levels—required an enormous feat: a civil engineering project dwarfing anything humans have ever attempted.

Extracting carbon from the atmosphere—essentially unburning all the fossil fuels that have ever been burned—carried with it an unfathomably huge economic and societal cost, potentially consuming between 10 and 50 percent of all global economic output by the end of the century. Most of the technologies involved plants. Photosynthesis is the cheapest, most effective, and most ubiquitous technology we know of to pull carbon out of the atmosphere: trees, tallgrass prairies, algae—

the restoration of natural ecosystems. Soils are still, by far, the least understood part of the global carbon cycle, and it's difficult to get good estimates on how much more they can be coaxed into holding. But the idea of carbon farming, nurturing plants to store carbon in the soil by managing the ratio of fungi to bacteria or intercropping multiple perennial species or by dozens of other methods, was so promising that we poured money into it.

Carbon sequestration became the main livelihood for hundreds of millions of carbon farmers around the world. Governments paid citizens to produce food, but to do so in a way that reduced climate change at the same time.

In an effort to maintain a path forward for outdated forms of air travel, billionaires (the few that still existed) pushed for enormous investment in carbon-negative biofuels on a continent-size scale. Another idea was to flood vast swaths of the world's deserts, seed the ponds with genetically engineered algae, and let the plants grow as much as possible. Another controversial method, called methane oxidation, would have involved intentionally releasing huge amounts of carbon dioxide into the atmosphere through chemical reactions in an array of giant fans—but would remove virtually all anthropogenic methane in the process, which, pound-for-pound, has an eighty-four-times-greater warming effect than CO2.

Throughout the 2040s, however, one primary negative-emissions idea dominated the conversation: bioenergy carbon capture and storage (BECCS), which involves growing plants, burning them for fuel, and capturing the emissions. BECCS, if scaled up to its full potential, would require an enormous expansion of oil seed farms to such an extent that it might ultimately threaten the world's food supply.

At a 2045 summit in Paris, thirty years after the original Paris climate accords, these proposals were soundly rejected. A window emerged for what UCLA geoengineering researcher Holly Jean Buck has called "a radically utopian way of removing carbon from the atmosphere." Using seized assets of long-bankrupt fossil fuel industries, governments began to coordinate a large-scale effort at carbon capture and storage in huge oil fields, fracking wells, and abandoned coal mines. By concentrating and capturing streams of atmospheric carbon dioxide and converting them into geologically stable liquids and solids—basically fake oil and coal—we began to run time backward.

The psychology of carbon removal, of erasing the sins of previous generations, is profound. But the reality is that there is no easy solution to undoing the current state of the world. For at least the next few decades we will endure a planet that's growing dangerously hotter every year. Embracing that cruel truth—and not running from it—will allow us to best ensure not only survival but a good life for as many people as possible during this era of fundamental transition.

## 2046–2049: HUMANITARIAN GEOENGINEERING

Even as the world switched to net-negative carbon emissions, global temperatures were still dangerously high and would remain so for decades or even centuries without further action. We began to wonder: What would happen if carbon removal failed to reach the scale necessary to avert further ice sheet collapses? Could we modify the weather—intentionally—to suit our needs? Could we create a planetary thermostat and turn the heat down? The complicated ethics of geoengineering suddenly came into sharper focus.

In 2017, the Red Cross convened the first-ever conference on "humanitarian" geoengineering. Such a concept, I'm sure, seems laughably naive to people who understand even the basics about the multi-millennial history of human power relations. Expecting that a planet-scale air conditioner could work, and could be fairly governed by all nations on Earth, is in bold defiance of our history with managing the proliferation of technology since the Stone Age.

The Red Cross's job, though, is to plan for emergencies. And, as evidence that our climate emergency disproportionately affects the people who have contributed to its causes the least, that's exactly what the Red Cross did in 2017. The ethical and responsible approach, they concluded, might be a limited-scale geoengineering effort that prioritizes the well-being of folks at the front lines. By the late 2040s, we had proved to ourselves that we were able to overcome existential challenges by working together, and so we decided to take on the prospect of cooling the planet.

The potential for deliberate large-scale intervention in the Earth's climate system has major implications in terms of impacts on the most vulnerable. Early engagement by the humanitarian community could have huge influence on how (and whether) geoengineering projects happen. Those who will suffer the worst outcomes need to be leading the discussions, especially given the plausibility of "predatory geoengineering," where reckless self-concerned actions may result in intentionally harmful consequences to others.

Pablo Suarez, who represented the Red Cross at the meeting back in 2017, spoke forcefully against the "potentially delusional assumptions of rationality" that have dominated discussions about geoengineering so far. The reason geoengineering is appealing, he said, is because it's a cheap, quick fix instead of doing the hard work of reorganizing society to prevent things like climate change from happening in the first place. "Climate change is the unwanted side effect of development," Suarez said. "No one likes to be the rat in someone else's laboratory."

Suarez posed the ethical decisions that a world on the brink of embarking on a geoengineering project would face: "If a hundred countries are better off, but Gabon is worse off, what happens to Gabon? Extreme weather will still happen, it will just be rearranged. Who will pay for humanitarian work in a geoengineered world?" There is no regulatory framework in place to help make these decisions. "We as humanitarians have a mandate to anticipate what could go wrong. And things could go wrong."

The risks of geoengineering were clear, even as early as 2017, when the Red Cross considered its long-term consequences. What was still up for debate, though, thirty years later, were its benefits. Spread in a thin layer in the upper atmosphere, sulfate aerosol particles induced a slight cooling effect, which limited the amount of heat and light the planet's surface received. Models that analyzed these scenarios continued to show that this would likely cause droughts and stunt the growth of plants. They also showed it would simultaneously reduce the energy available for extreme storms and reduce the intensity of rainstorms. It might also delay by centuries the terminal collapse of the rest of the glaciers in Greenland and Antarctica.

Geoengineering the planet, Suarez warned in 2017, would be embarking on a plan of planetary chemotherapy: injecting a harmful substance into the atmosphere to try to undo past wrongs with a full understanding that there may be serious side effects. "In extreme cases," said Suarez, "those side effects could kill the patient.

"Is that death different than the death caused by cancer? Maybe, maybe not. Is the suffering different? Maybe, maybe not. We know we are confronting difficult choices."

Still, this was one of many choices we had to face, despite our decades of hard work. What would happen if the global humanitarian community embarked on a limited-scale geoengineering project late in the 2040s? What would happen if people, thrust dangerously to the practical point of genocide, embarked on a plan to demand atmospheric justice?

* * *

The worst-case scenario isn't necessarily that we'd start a geoengineering project but that we'd suddenly stop it.

An abruptly halted geoengineering project would have shock

effects even worse than the relatively slower, but still disastrous, phase-out of aerosols. A sharp rise in temperatures would produce an overnight shift in weather patterns that might prove catastrophic for people and species already stressed by decades of dangerous climate change. Far-ranging ecological consequences would essentially be irreversible, accelerating a mass extinction.

"The best-case scenario is that we don't have to deal with this," Holly Jean Buck told me. But if we do decide that a geoengineering project is absolutely necessary, "the best case is a very short solar geoengineering intervention that is small in scale and doesn't use very many aerosols, and keeps temperatures down while the world is seriously pursuing carbon removal, and wraps up within the century."

A slow, managed phase-out of geoengineering over a time span of many decades may sound great, but when have humans ever been able to plan something that actually works as intended on a timescale of half a century?

More likely, Buck told me, there'd be a nasty overlap between solar geoengineering and authoritarianism. She outlined a nightmare scenario: "What if you have an authoritarian leader, and they say they're going to fix everything with solar geoengineering, and it appears to work, and then they use that as a justification to continue being in power?"

Another scenario terrifies Buck: What if a dying oil industry decides to pivot to carbon capture to turn a profit off an escalating global crisis? That scenario alone is enough reason to consider nationalizing the oil companies, she said.

In the world where we're already carbon neutral and we're considering using geoengineering as a humanitarian supplement to ease our transition back to a stable world, as well as a conservation tool to help relieve stress on ecosystems and prevent species from going extinct, it may not be such a bad idea. "There's basically zero research on [using geoengineering like] this," Buck told me, "and it seems to me to be one of the more important justifications for considering using it." That world, obviously, is far from where we are now. But as this book has hopefully shown, it's not so far-fetched.

If political and social responses to climate change fail—or if we are so wildly successful that we expunge all aerosols from the sky and bring about a rapid rise in warming—would we maintain the same courage and optimism to, once again, trust artificial planet-wide cooling technology to save the day?

On the surface, geoengineering seems like an uncompromising escalation of the problems that got us into the mess we're in. Still, it's worth remembering: we've already massively changed the planet's atmosphere and ecosystems, to an extent only rarely seen in all of Earth's history, barring meteor strikes. Would it be worth attempting to explicitly undo some of that damage? On the other hand, intentionally modifying the planet's atmosphere would be adding a whole layer of complexity to a world in the midst of rapid change.

In the future, "we" will need to decide what "humanity" means and what "the Earth" means in an era when the fates of people and the planet are more intertwined than in the entirety of history and the responsibility for this crisis does not sit equally.

What we do to the Earth, we do to ourselves. Our best hope to avoid geoengineering might be mutual aid, trust, and solidarity on an unprecedented scale.

In 2049, at a global summit convened by the Marshall Islands, Indigenous people from the Arctic to the Okavango to the Australian Outback debated what to do in a months-long conversation that convened representatives from universities, religious groups, and the thriving youth ecological societies. The result was overwhelming: we would attempt to decolonize the atmosphere, very slowly, in an attempt to return the sky to those who were still alive and the hundreds of generations who were still to come. It was a centuries-long project that would honor the incredible spirit of cooperation humanity had developed. It was a way to repair the harm caused to all non-human species that had suffered for so long.

By 2050, we had accomplished the bulk of the work we needed to do as a civilization to stabilize our climate and ensure a livable world for the countless generations that would follow. This achievement, derided as wishful thinking or impossible just a few decades earlier, proved to be the most significant, most heroic, most improbable revolution in the history of the world.

The tech fixes continued: industrial-scale carbon capture, limited geoengineering, artificially intelligent management of variable wind and solar resources. But mainstream culture shifted from unregulated growth at all costs toward a mutual flourishing of people and nature, in which the circular, ecological economy created vibrant places for work and play that complemented their surrounding environment, not worked against it. We moved forward by prioritizing all kinds of knowledge, not just techno-utopianism.

The world changed so much and so quickly that we could no longer afford to ignore or rule out any reasonable way to stabilize the planet.

In these past three decades, not everyone has been on board with all these changes. There are rogue countries, rogue industries, holdouts from the past, and war may still be unavoidable. But mainstream culture has shifted so much that we have developed systems to hold these people accountable and prevent them from harming the rest of us.

Through the first half of the twenty-first century, we came together and grappled with ongoing rapid climate change and what it meant to be alive at a moment when extinction was a reality and the rootedness of place was not so permanent as we once believed. To meet this need, we developed and practiced new spiritualities, which granted us the patience and grace to start the second half of the century with a carbon-neutral global society and a circular economy that aims to repair centuries of colonization and exploitation. We've fostered a rejuvenated new ethos for humanity.

We learned that our lives, our prosperity, our cultures, and the existences of the animals and plants we share this world with were bound up in the interconnected web of rain, snow, sunshine, winds, and weather that make up our atmosphere. By destabilizing this balance we learned that we were destabilizing what literally makes life possible on our planet. Now that we've stabilized our relationships with one another and with the Earth, there's no limit to what we can imagine.

## TAKING CARE OF THE EARTH IS A NEW (OLD) SPIRITUALITY

Since the dawn of time, humans have understood the importance of recognizing that we are one with the planet.

There's no way to put into words the softness of a sunset while sitting on a cliff high above the California shoreline, waves crashing in the distance, while pondering the sheer vastness of the Pacific Ocean before you. There's no way to describe the taste of a peach. There's no way to perfectly capture the giggle of a toddler as they tickle a caterpillar, or the wonder of a conversation with someone who has walked a different path than you but is intimately connected in that shared and fleeting moment. These simple luxuries ground us to the planet that brought us into being. There's a reason we feel deeply at home when we share these simple joys with one another.

What makes us human? What makes us love one another? These are the things we're fighting for when we're fighting for a stable climate.

By 2050, almost everything we know about the world will have changed. But these core, shared experiences of humanity will remain. The greatest gift we can give our future selves and those we share this space with is to radically act in our current moment with the deep and transcendent love of visionaries.

Kyle Whyte's idea of a civilization based on kinship, which he articulated in 2019, has a transformative effect, if you consider how it would radically change our idea of self, and what we might devote our lives to hoping for and working toward. "Instead of trying to force climate change solutions onto people," he told me, "if we're more attentive to relationships and consents, that kind of system would suggest a very different type of future than the one that I think a lot of folks are envisioning."

The urgency and anxiety that people have started to feel as climate disasters become more and more obvious is rooted in this reluctance to radically change a system that has benefited us. But if we take Whyte's vision, it will be easy to recognize that urgency as an illusion manufactured by a system built on distrust, blame, and hyper-individualism. In a world where we devote substantial time to building trust with one another—a relationship-based civilization—that urgency and anxiety will melt away because we will know, deep in our bones, that we are all looking out for one another. That we'll all be cared for, because finally we are working with the planet, not against it, just as Indigenous people have done for thousands of years.

Mary Annaïse Heglar, whose writing has come to exemplify the brave kind of visionary leadership that our moment demands, said that the world starts by imagining something. "Climate change isn't something that's passively happening," she told me, "it's an intentional act. Doing nothing is doing something. Burning fossil fuels is an action."

In 2050, science might also look and feel very different from today.

"Science is how we got climate change," Heglar told me. "It took science to figure out that fossil fuels can be used for energy. It took science to figure out where the fossil fuels were located.

"If we decide that we're going to use science for constructive purposes and in harmony with the planet and we start to listen to groups of people who have never lost their relationship to the Earth, then we start to see a more benevolent science, a more holistic science. Because what is science if it's not the study of the Earth? If Indigenous people are seen as working in tandem with scientists, I think that would be a really beautiful thing. If we didn't have all of these ready-made barriers to separate ourselves, how much more innovative could we be? How much more empathetic could we be?"

From an Indigenous perspective, our civilization in 2050 might not even be human-centric at all.

"We should really start to take a back-seat role that centers us as stewards," said Kelsey Leonard, the ocean policy advocate from the Shinnecock Indian Nation, "as people who are responsible for fostering connection to our other non-human relations on this planet and allow for that life to be prioritized over our own.

"If we can really start to see the principles of UNDRIP [United Nations Declaration on the Rights of Indigenous Peoples] enshrined and applied to climate change policy and how we envision a climate conscious feature that's sustainable and that has equity, it needs to include those key aspects. If we actually lived up to the principles enshrined within the document, we'd have a better planet, a safer planet, a healthier planet, not just for Indigenous peoples but for everyone."

* * *

It's 2050. The world is carbon neutral. The economy is circular. Society has transformed. Our world is a place that has decided to radically change in its entirety because places like the Marshall Islands matter. Because you matter. Because we couldn't just go on like we had anymore.

Inequality still exists, it always will. But humanity has learned that we share more with one another—and with every other living thing on the planet—than we had ever imagined.

Our brush with catastrophe brought us into a painfully beautiful planetary relationship. It is such a remarkable time to be alive.

# EPILOGUE

We are not all equally to blame for climate change. Yes, the Amazon rainforest is burning at a record rate. Yes, sea ice is at an all-time low. Yes, genocide is ongoing, because rich people are trying to get richer.

This is the planetary dystopia of our time. The anger we feel toward the state of the world right now is worthy and necessary. It helps us focus our action.

But that's not the only story we can tell. We can also tell stories of love. Those of us with power must fight for justice, for a new system that values survival over profit and flourishing transformational change over the status quo. This is radical hope: knowing that the future can be better and knowing that we are the ones who must make that future happen.

The biggest change is within our own minds and hearts, to envision that a world like this is even possible.

This moment is scary, and traumatic. But we are in it together.

In all these different possibilities presented in this book, the most important one I want you to consider is that our future is all about the narrative that you tell yourself. That's literally how we are unconsciously able to move throughout our day, by trusting that the actions we take will lead to specific outcomes. Working for a good future makes that future possible. And working for a good future can't happen if we don't believe it's possible.

## WHAT'S THE MOST IMPORTANT THING YOU CAN DO?

My theory of change is simple: billions of people just showing up in their own lives, energetic and ready to struggle together.

Too often we hear that in a representative democracy, the most important thing you can do is vote. But what about the other days of the year? This is a crisis; we don't have time to just wait until the next election.

To match the scale of the climate emergency, we need radical societal change. This always intimidates people, because it sounds impossible, too big to wrap our arms around. But what if societal change is just a bunch of individuals living their individually radical lives? This includes you. If we start with changing our own behavior, we can effect larger societal change. Because we need to accomplish both the radical systemic change we seek and radical personal changes in our own lives.

At first, this will feel uncomfortable—but will it feel more uncomfortable than dooming yourself and everyone you love to a future with a smoldering remnant husk of a planet?

So which individual actions matter most? The honest answer is, the ones that help make you personally more connected to the world and everyone in it. You should definitely run all the aspects of your life through a carbon-footprint calculator (you'll probably be surprised what you find), but the point here is to re-imagine your relationship with the world in a way that helps you live a more fulfilling and healthier life to ensure our ecosystem survives, not to check off a certain number of boxes.

That's why I think the single most important thing each of us can do about climate change is to talk about it. With anyone who will listen. We exist in a crisis, and during a crisis, there are voices that often get drowned out. When someone talks with you about the climate emergency and how it's affecting them in their life, listen.

Talking about climate is what builds social and political pressure for radical action. It's also a radical personal step itself in a world that treats conversations about climate change as taboo, even in the company of friends. Talking about climate change is how we build a better world. Learning and listening and getting excited about new ways of existing on our beautiful planet is impossible without conversation.

What about the hundreds of corporations that are responsible for the bulk of emissions? What about the oil and gas industry? We need to nationalize these horrible companies or shut them down. But to do that, we need to talk about it first.

And that means setting the idea of apocalypse aside. The old world is already gone. We will remain lost, floating, undefined until we start our shared work on building the new one.

One month after Hurricane Dorian's apocalyptic wrath through the Bahamas in late 2019, more than a thousand people were still missing and seventy thousand were homeless—in a country of four hundred thousand. The storm also inflicted a severe economic shock: 58 percent of the country's GDP was lost overnight. Prime Minister Hubert Minnis called it "one of the greatest national crises in our country's history."

But even these bleak facts fail to convey what it must have been like for the people hunkered down on that dreadful day. Unearthly sounds of rending sheet metal and splintering wood, the rain and sand and ocean water pummeling every surface. Stories emerged of a man, wheelchair-bound and sitting in floodwaters for forty-eight hours.

It is a testament to the human spirit that anyone survived at all.

As a force of nature, Dorian was beyond compare. It was the strongest hurricane to make landfall in all 150 years of documented storms in the Atlantic Ocean. But its cause was anything but natural. Centuries of decisions were embedded in those winds and waves.

In the immediate aftermath of the hurricane, headlines described the Caribbean archipelago with words such as "crippled" and "hell" and "devastation"—and then, as usual, the international media went almost entirely silent.

Climate change itself is simple. I can explain it in one paragraph: by a quirk of physics, fossil fuels are an almost perfect store for energy, and their discovery helped accelerate centuries of colonialism, locking us into an extractive relationship with our planet and one another. The subsequent imbalance in resources was exploited by those with economic or military power, enriching the few at the expense of the many.

But the fix is not simply technical. The too-familiar apocalypse narrative leaves no room for justice or regeneration. We must do better. Somehow we must also learn to treat one another better.

But how do we do that? Figuring out exactly the steps we must take to address this emergency, that's hard. There are no computer models, no satellites, no radar systems, no hockey-stick graphs that can help us chart our way toward the kind of civilizations we urgently need to build.

Surely the only way to begin is to reckon with the gravity of this moment, reconnect with our shared humanity, and forge on together. In the words of President John F. Kennedy, we must do these things "not because they are easy, but because they are hard." Slowing down planetary collapse is the hardest thing we may ever have to do as a species, but it is also—unequivocally—the most important.

In all of history, no other human force besides armed conflict has driven more people from their homes. No other force could destroy an entire island overnight, could make vast swaths of the world uninhabitable, or cause the graves of hundreds of centuries of ancestors to permanently sink beneath the sea.

Today there are more than 70 million forcibly displaced people around the world. There are no reliable figures on how many of these displacements are related to environmental degradation or climate change, not least because climate change now deeply affects almost every place on Earth. In our lifetimes, without a radical change, this number could increase tenfold.

Climate writers often slip into a war metaphor. But climate change is not a war. It is genocide. It is domination. It is extinction. It is the most recent manifestation of how powerful men throughout history have sought to steal from the less powerful and dismiss them as merely inconvenient. Understanding climate change in this way transforms everything.

Worse than the way we talk about this moment in history is the realization that the narrative of climate apocalypse is not a catalyst to action. Instead, it helps reinforce the business-as-usual trope: If we're going to lose the Bahamas anyway, why change course?

From the descendants of slaves in the Bahamas, forced once again onto boats as they fled, to the burning homelands of Indigenous peoples in the Amazon, their forests cleared for cattle ranching and soy plantations, to the Syrian refugees of a conflict in part triggered by years of drought, the climate emergency looks like violence.

There is no need to convince anyone in the Bahamas that everything is different now. As a country that has contributed a mere 0.01 percent to global greenhouse gas emissions yet suffers some of the worst consequences of the climate emergency, no one there needs convincing that inequality is part of the problem.

What they also know with certainty is that it will take years—

decades, even—for the residents of Abaco and Grand Bahama islands to rebuild their lives from the rubble. Writing in *The New Yorker*, Bahamian writer Bernard Ferguson captured this feeling when he said: "The death toll, when tallied, may never be a complete or accurate expression of the lives that the storm claimed."

For people living on the front lines of climate change around the world, there is no physical defense against this kind of unnatural, human-made violence. "The most potent defense that we have," added Ferguson, "is to strategize and organize collectively, across countries, to reverse our course."

This is what the climate emergency looks like—not stories of solar tech and world leaders signing a lukewarm, lowest-common-denominator agreement, and definitely not a simple statement of long-established physical science.

It is the minute-by-minute revolutions happening in nearly every home and neighborhood around the world where people are simply claiming the right to exist. It is not just the contemporary image of a family standing amid their island ruins; the climate emergency looks also like the five-hundred-year history of colonialism in the Americas. This has been happening for a long time, because climate change is a crisis of our relationship with one another and with nature.

What this moment needs, more than anything, is moral clarity, the kind demonstrated at the United Nations by a Swedish teenager and countless other young people from around the world.

We need to know, viscerally, that we can no longer abandon our neighbors in their time of greatest need. We need to relearn our interdependence. We need to learn a way to rewrite this story that doesn't end in apocalypse.

* * *

In the meantime, my most immediate advice is to go outside and enjoy your present Earth. There are physical and mental benefits of getting outdoors. Do a bunch of these things (or at least a few of them):

**Go for a walk in the forest.**

**Make art (outside).**

**Go snorkeling.**

**Actually meet another living person with shared interests.**

**Look at the bugs.**

**Go bird-watching.**

**Go to a star party and ponder your place in the universe.**

**Go kayaking.**

**Hike across an island.**

**Go to an orchard and pick fruit at peak season.**

**Stay in a tree house.**

**Go to a baseball game.**

Doing these things will give you inspiration to do more things:

**Open a (nonprofit) business:** As the owner of a house-cleaning business that uses green chemicals, for example, you have a chance to talk about climate change with each of your customers.

**Run for office:** For the past few years, you've heard firsthand from members of your community demanding change. You can help bring it to them.

**Become a teacher:** The best way to change the future is to help the people of the future empower themselves through knowledge.

**Become a farmer:** Whether you live in an urban, suburban, or rural community, the land is what sustains us. You owe it to the Earth to be a good steward of the land. Find a patch of soil you can farm and enjoy its many bounties.

**Demand local action:** Working with local community boards on something as simple and necessary as adding more bike lanes will help ensure a people-powered future comes into existence. And because bikes are one of the most energy-efficient inventions in human history, you'll immediately improve the environment.

Most important, talk about climate change with people you are close to. Build solidarity and like-minded support networks and a shared vision of a better world.

## HOW TO TALK ABOUT CLIMATE CHANGE WITH PEOPLE YOU DISAGREE WITH

I grew up in a coal-mining and farming town that voted for Trump by a 43 percent margin. Talking to my parents about climate change has always been difficult.

But the time to be angry about climate denial is over. It's time to find common ground—there's literally no time left to debate the problem anymore.

By building up an irresistible vision of a better world, we'll be able to overcome the insidious influence of the fossil fuel industry with the biggest, most powerful people-centered movement in world history. That kind of a revolution, to me, is the only path that will lead us to where we need to go in the time we have left.

One person I trust on things like this is Katharine Hayhoe, a climate scientist who has made it her personal mission to convince people that we are more alike than we think. She told me, "The reality is, we're never going to agree on a lot of things. I think we have way more in common that we can move forward. We have so much in common, and no policy should be put in place just for one single reason. We can't really afford that luxury. If somebody's on board for a different reason, that's okay! We need to stop being all purist about it."

As an example, Hayhoe points to the issue of coal as one that can bring different groups together. "We shouldn't be using coal, because it kills people. Two hundred thousand people in the United States and five and a half million people around the world die every year from burning coal and from burning gasoline. If you're a person who is simply concerned about that, but you say it didn't cause climate change, I'm totally fine with us being on the same team advocating for a price on carbon or stronger emission controls on fossil fuels."

Her words remind me to be patient during times when I want to just burn down the world. Coal miners are not the enemy. Your cousin who flies business class isn't the enemy. Your neighbor who eats meat is not the enemy. The enemy is the system we're all embedded in—the same system that's been the engine of extractive, colonial, genocidal exploitation of the only planet we all have.

Even though this book tries to outline how a different system might come into being and what it might look like, I'm the first to admit that I'm not sure what exactly is going to happen during these next critical decades. The good news is, we get to envision and enact this future together. And I have no doubt we'll do it—because we have no other choice.

## WHAT WILL YOUR WORLD LOOK LIKE?

This moment in history needs you. Knowing what you can personally do is easier than it seems. You already know what issues you're passionate about. You already know what kinds of things you're good at. Where those overlap is where you should devote the rest of your life.

Your life is your own story to write. You get to decide—every single day—what your relationship with the world is. That's the beauty of this moment: you have the ability to act much more quickly than the political system. In fact, you could even change some relatively major stuff about your life right now! Seriously: put down this book, go for a walk outside, and think about it. Imagine a better world and how you might be able to play a part in it. Invite a few friends over to talk about it. And then figure out how to make it happen. You can do it. I believe in you.

* * *

To help you figure out how to make your future Earth a reality, I've worked with some friends to prepare the following action guide and reflection exercises to inspire your imagination to contemplate what's still possible in the age of warming.

## A GRIEF EXERCISE

Grief is more than sadness about loss; it is a bodily experience. We often think about it in the context of the death of a loved one, but grief can extend to other losses. As climate change is featured more in the media and as we are bombarded with images of ecological loss, many people may be working through unconscious and unprocessed grief. Building on listening practices, we can begin to process and metabolize our pain about the Earth with a creative mourning ritual, which can help us move some of the heavy energy around this topic into a more creative, communicative opening.

Listening and care are two acts that can look like "doing nothing." Anyone participating in these activities, however, knows how much work it is to undo some of the cultural lessons we may have picked up about hyperproductivity and the cultural mandate to push past natural boundaries to produce endless growth.

It can be helpful to announce that the purpose of gathering is not to "solve" climate change. In my experience, people often become very solutions-focused when discussing climate change. This can put a lot of pressure on the conversation, and it can also be derailed by a hyperfocus on a single solution. This focus can often be a way to escape the discomfort of being with this thorny and often upsetting topic. In order to make space for people to explore and express their feelings about climate change, and to start the difficult transformation of personal grief into collective action, deep listening and care work is important.

## GETTING GROUNDED

People tend to seek refuge in a storm. By grounding into our experience, cultivating an inner sense of safety, and locating some resources we can turn to, we can make space for one another to experience the challenging aspects of grappling with the rupture, loss, and grief of ecological crisis.

Sometimes when we come to group practices we might be bringing a lot of energy from whatever we were doing before, or we might be thinking about what we have to do next. This grounding practice can bring your awareness into the actual time and space you are in and begin to settle your central nervous system. Of course, sometimes when we become more aware of our experience, we become aware of just how tired, anxious, or hungry we are. Whatever you notice is okay! The point of this is to start unwinding some of our forward momentum so we can all be in the same space together.

### Step 1. Assume an attitude of receptive curiosity

This practice is more investigative than goal-oriented. Part of becoming familiar with your own subjective experience and its relationship to your nervous system is just being curious about what it feels like. This is not something to "get right." If you don't feel more calm when you're done, it doesn't mean you did it wrong. You're learning about your experience. Maybe for you, there is numbness in part of your body when certain thoughts come up. Maybe another part of your body feels more engaged. If, a little at a time, we are able to start noticing and not judging these things, then we stand a chance of offering those aspects of our experience some compassion.

**Step 2. Start by just sitting**

Give yourself permission just to sit. You don't need to make anything happen or achieve a state of extreme focus. If you find yourself defaulting to a state of striving or straining, give yourself a break. Knowing that while you might find yourself going on elaborate list-making adventures (it's always grocery lists for me!), you don't *have* to indulge every idea or thought that comes up. You can just sit with them, neither entertaining them nor pushing them away.

**Step 3. Tuning in to the senses**

Part of tapping into our body's innate resilience means being able to drop out of the analytical aspect of the mind into the physical felt sense of consciousness. For some people who have suffered trauma, the body does not always feel like a safe place to be, so please feel free to try this for only a minute at a time. There is no need to force yourself through this. We explore, and learn, and can take whatever small steps we can when we have the resources to try.

A great place to start is with the felt sensation of your feet against the floor. Notice if you start thinking or narrating about your feet, and see if you can bring your awareness back to the felt sensation. Feel the gravity. Feel the pressure. What is interesting about this? Does the floor feel solid? Can you imagine that you are supported by and connected to the floor? Continue to notice the other points of connection. Your legs and butt against the cushion or chair. Be curious. Breathe. Scan your back from the base of the spine to the shoulders. Are there positive sensations? Points of tension? Can you get curious even about the neutral sensations? What is the temperature? Just be aware, and be curious. If your back is against the seat-back, notice that sense of connection there. It's okay if it's just a vague or general sense of sitting. Allow yourself to feel it without forcing yourself to concentrate.

If, by the conclusion of this, you were able to feel into your body for only a moment, that's still awesome. The process of familiarizing yourself with the felt sense is not easy for many of us, whether because of trauma or because our jobs require us to be disembodied on the internet all day. Giving yourself even a brief opportunity to feel with no particular goal can be incredibly healing. Even just cutting through the forward, task-based do-do-do momentum of your day can be restorative.

**Step 4. Checking in with the ground**

Part of a grounding practice is learning to be in the space where we are. Is there anything about the particular moment or room that you are in that points to this specificity of place and time? Sounds, temperatures, a general ambience? These are not things we need to push out of our experience in order to concentrate.

Rather, we treat all aspects of where we are and use them. You might notice that you are straining to be elsewhere or to project yourself into the future or past. This tendency is totally fine. We're just going to see if we can use the senses to come back to the present. The breath can be a great anchor, or the feeling of connection with the floor. Maybe there is a droning sound from an air conditioner. That can also serve as a place to rest awareness.

**Some questions you can journal about after doing a grounding practice:**

- What was interesting about sitting with the felt sense?
- Were you able to notice positive sensations? What were they? How about negative sensations? Were you able to feel neutral sensations? Was anything interesting about any of these?
- If you were able to feel very in the room you are in, what were the characteristics of that feeling?
- If you were able to notice the mind imagining the future or reminiscing about past experience, what was that like?

## REFUGE

We all have inner resources that we can turn to when our experience of the world feels challenging and stressful. Consciously assembling some of those tools can be a great tool for this work. Here are some tips for cultivating the resources that can help guide you through this work.

1. Who are some people you could call on to guide and inspire you as you confront your feelings about climate change? They could be ancestors in your family who lived through difficult events, historical figures, mentors, or loved ones who give you a sense of being cared for. They could even be characters from literature or film who embody the resilience and values you might hope to embody. Take some time to call these beings in as your resilience team.

2. Is there a place where you feel particularly safe and at ease? Take some time to call that place to mind as a sense experience. What does this place smell like? What are the sights you can see? Are there physical sensations associated with this place? What does it sound like? Take some time to make these sensations palpable. Let yourself experience what it is like to be there. And let your body feel the sense of ease or safety, or whatever other feelings come with letting yourself be in this place.

3. If this place and this team were to live in a place in your body that you could access when you needed them, where would that be? Take some time to see what comes up. It can be easy to use your heart or the center of your chest, with a gesture of placing a hand there, when you need to call up your place of safety and your resilience team. But something else might come up for you. Experiment with some movement and listening to your body, and explore where this safety and resilience could live.

4. Make a commitment to remember this sense of ease when you feel unsafe, uncomfortable, or agitated during practice or conversation. This isn't about pretending you don't have challeng-

ing emotions about climate change, but rather enabling you to be present and to offer refuge to yourself and others even as inner or outer turbulence arises.

## ENGAGING BEWILDERMENT

What happens when you have no frame of reference? You're more open and you notice more. Once you "know" and your mind has related your present experience to preconceived frameworks, you only notice things in a way that relates to that frame you've chosen.

The mind is always telling stories about our experience, and if we notice when it's doing that, maybe we can learn to tell stories in an open and generative way rather than a nihilistic and despairing way.

Engaging Bewilderment as a practice can be experienced in a number of ways. Some methods I've practiced are playing recordings with unusual instruments or field recordings with a variety of sounds; participants listen and notice when their mind tries to identify what is creating a sound and see if they can open their awareness back up to the full experience of listening without focusing on what is producing the sound.

Another practice is to have a big bag full of items with a variety of textures. Participants close their eyes and reach into the bag to select an item. It might last for only a split second, but can they just feel the item without rapidly identifying what it is?

After this, participants can spend ten minutes sitting and

journaling about what came up for them in this exercise, any insights they had. You can then open it up to a discussion about the experience.

## LISTENING PARTNERS

Bearing witness to our own suffering and the suffering of others is a form of compassionate action. Breaking into pairs, we take turns listening to each other. Each person has five minutes to speak. The speaker is practicing putting their felt sense into words. The listener is practicing listening, without shaking their head yes or no, and without planning their response. If the listener notices their mind wandering, they can simply bring their attention gently back to their partner's spoken words. After five minutes, the partners switch. After each partner has spoken, the final five minutes are spent in open conversation between the two, where they can discuss what it felt like to do the practice. The group then reconvenes and people can discuss any insights they had doing this practice.

## PARTICIPATORY FRAMEWORK—CREATING A GRIEF RITUAL

Building on the practices of being open to bewilderment and deeply listening to and holding our own emotions and others' perspectives about the climate crisis, we move now toward metabolizing our grief through collective practice. How do we as individuals process rupture, loss, and change? As the stress of change and loss impacts our bodies and minds, we might react in the form of patterns we learned at a young age. Depending on how we were taught (or not taught) to navigate ruptures in our lives, we might be repeating these patterns in response to the climate crisis. Some of these reactions might have helped us survive as young people and maybe made our suffering easier to manage for ourselves and those around us. Our work now is to sort through and process some of these patterns so we can move toward responding and participating rather than reacting from our survival conditioning. Responding is more flexible, intentional, spontaneous, and curious. We can face change together rather than reacting in ways that prevent us from truly being with one another. There is a lot of pressure in our culture to "get it together" and "move on" from loss without the support to really process it. Here are some tools a group can use to do some relational healing around the climate crisis.

## Acknowledging the Loss

Being able to speak the truth without the pressure to sugarcoat our experience of loss is an important aspect of acknowledging the loss. We cannot begin to create a coherent story for navigating the climate crisis without first being able to name what we are experiencing. Here are some things to discuss in pairs and in community, or through journaling:

- What do you feel you've already lost to climate change?
- What are you afraid of losing?
- Can you name the sensations and/or emotions that arise as you identify what feels different?

- What are some values you feel are being lost that you might like to embody—in the short term within this work, and in the longer term in your daily life?

## Re-membering

Re-membering, or piecing our story back together after the rupture of loss and change, is a form of bringing that which is felt to be absent or lost back into our present-moment experience. Sharing this with others allows a group to prepare the ground for cocreating new narratives. It also allows the group to return to life and interactions with fresh eyes. If we don't show up for one another when there is suffering and pain present, we run the risk of shallow stories emerging, rather than deeply rooted ones coming to fruition. Feelings of pain, anger, frustration, and despair are reactions based on care: we care about people, places, and things that are suffering and under threat from climate change. Can we honor that care as a form of life, love, and connection?

- What reminds you of the ways you have loved and cared about this world, other beings, the environment?
- Can you express this through poetry, drawing, sound, movement?
- Write a letter to yourself from your future self or an imagined future community. What advice or words of support do you feel moved to communicate?

## Integrating and Sharing Gifts

Based on all of the work so far, the integration and sharing can come in many forms. If people have been using visual art and drawing, you could experiment with a large shared mural on butcher paper, improvised from work people have done on their own. If people have been writing letters, poetry, or journaling, you could invite them to share. Movement and sound can also be shared and spontaneously combined if people feel so moved. Facilitators can act to encourage and support this cogeneration of collective expression. Some key things to explore:

- What feels like a major insight or breakthrough that you and/or the group have had through this process?
- Is there something that came up for you during this process that you would like to bring with you out into your life? A small step that you might actually take? Or a feeling that could serve you in difficult times?
- Is there anything you feel moved to share with an individual or with the group?

### Best Practices

Facilitators should have some experience holding space for collective emotions, whether through facilitating support groups, meditation circles, as a caregiver, or as someone who has consciously moved through their own grief about interpersonal loss. Also helpful is any background in trauma sensitivity and movement, such as yoga or other somatic practice. If you can find a collection of people who each have some experience in one of these categories, the group can also function as a kind of skill-share. I think of this group as a sort of "resilience team" that can help hold space for newcomers to the topic for the difficult emotions that often come up when discussing change.

Having some background in grief or trauma sensitivity can be great for work like this, but if you're a newcomer to this work, don't let that stop you from offering compassion and care for fear of "doing it wrong."

**Emotional First Aid.** Many people come to this work with a lot of fear and pain. Even as facilitators, we might notice these feelings come up for us. While we want to leave some space for working with this pain, it's also critical to be aware of psychological and spiritual injury that can occur when people feel rejected, lonely, and panicked. When we are creating a space for working with something as potentially upsetting as climate change, we need to have some awareness of these triggers. Avoid criticizing the way anyone responds to the climate crisis, and aim toward curiosity and more questions.

**Speaking from your perspective.** Situating your words from your perspective can be a great way to cut through assumptions and biases. When you speak for yourself rather than dictate what an assumed shared experience is, you offer an invitation for connection, instead of universalizing your opinion. When we speak from our own perspective, we can leave space for this perspective to change with awareness and in relationship to the group. This can include starting sentences with "My experience has been," or "I have noticed," rather than an assumed "We always . . ."

**Non-judgment.** There can be a tendency to label some emotions good and some negative, and then to disown or avoid those considered negative. In this work, we can take the perspective that all emotions are valuable information from our bodies and psyches in relationship to our environment. Be careful not to comfort people out of their challenging emotions, but also be careful of overindulging any emotions. We aim to strike a balance by being engaged witnesses offering refuge and serving as a container for experiencing this process together.

## AN IMAGINATION EXERCISE

Welcome.

We are in a moment of apocalypse. We cannot return to the world that was, because that world no longer exists. Instead, it is up to us to help bring a new world into being. People all over the planet have been imagining beyond apocalypse forever. In your blood lineage, there is someone, somewhere, sometime in recent or distant history who has done just this. Step one is to remember that you can do this—that we can do this—because we have done it before.

The following pages are a guide and an invitation to imagine and enact the story of your future. Imagining how the world could be transformed is hard, emotional work. But, it is necessary. Doing this work with friends makes this hard work easier and less isolating.

The bad news is that, for many people, it too often feels easier to imagine the end of the world than to imagine a world/worlds beyond our current system of extraction and exploitation. The good news is that the power to change that reality is already inside you. The key to unlocking that capacity is grief work.

As you begin to grieve loss, you start to engage your imagination, your creativity, to foster personal agency and positive action for the future.

Remember, care work is climate work. We are not going to get to where we need to go by tinkering around the edges of a fundamentally unjust system. This is not just about building bike lanes and solar panels; this is about building a new society. We've got to know in our hearts that every single person truly matters, and that we *can* bring one another through this time to a future Earth that works for everyone.

Below are a set of principles that guide much of my visionary work, followed by a sample exercise for practicing local imagining work. Stories and draft agendas can be found at www.ericholthaus.com/futureearthstories.

## PRINCIPLES FOR IMAGINING A FUTURE EARTH

**"Small is good, small is all."** —*Emergent Strategy principle*
Start smaller than you think. If this is your first time doing group imagination work, three people (including you) is a good number.

**Begin with the end in mind.**
Ask yourself why you are doing this imagining work. Do you want to create some hope in your neighborhood? Do you want to catalyze your friends or family to take action? Are you trying to shake up conversations in your whole city? Whatever it is, get clear about it and let it inform your planning process. If you want to create hope in your neighborhood, then that goal should shape who you invite and your plan for stories you want to create

with the group. This goal will look very different than if you want to create and implement compelling big-scale reforms and new visions in your city.

**Collaboration over competition.**
In Western industrialized nations, we tend to believe competition is the best way forward. But the story of evolution is largely a story of collaboration. Visioning and working together are critical for survival. (And there are people who still believe competition is the best way forward. Be wary. But remember that everyone has the capacity to transform.)

**Iterate, iterate, iterate.**
There is no "right" vision. Imagination influences reality, which influences future imaginations into infinity. Create your vision, learn about other people's visions, let your visions influence one another, then vision again! Rinse, wash, repeat.

**Me to we.**
You matter. Your individual vision and imagination are necessary. *And* the power to spread and scale comes from weaving and building our imaginations into/onto one another. Never forget that you matter. Never forget that we matter. We can't get there without you, and you can't get there on your own.

**"Wherever there is a problem, there are already people acting on the problem in some fashion."** *—Allied Media Projects network principle*
Find them and listen to them. They might be in your neighborhood, in your watershed, or on another continent, but they exist.

They might be people you have previously ignored. They might be people American society has generally deemed less-than or people our social systems have actively marginalized. Listen anyway! Sometimes the best thing you can do to help solve a problem is to pass the microphone.

## SAMPLE PROCESS GUIDE

**Step 0.** *Gather your people.*

People: 3 to 8 people

Time: 2 to 3 hours

Location: A living room, or den, or dining room—any space where you won't be interrupted and can talk openly and at ease.

**Step 1.** *Check in.*

Go around the circle and let each person share how they're doing today, as well as what they're most excited about for this time together.

**Step 2.** *Set your scene.*

*Pick a place and time at which to dream.*

**Place:** Ideally it's a scale at which you *already* believe you have power to make change: homes, building(s), or neighborhoods are usually safe bets. The scale of a city is probably a stretch, but it might not be, depending on who's in your circle. At this stage, state or national scenes are almost definitely too large.

**Time:** *Five* years often feels far enough away to allow for some real change but close enough to feel real, not intangible. Or maybe you want to start small and focus on having a single conversation with a friend—five days should be enough for that, and depending on who it is, imagining that conversation in advance might be super helpful. But if that doesn't work for your group, figure out what feels better.

*Brainstorm some problems* that are being faced at the scale you picked. I suggest ten to twenty minutes for this. Then, as a group, pick one or two problems that you are collectively most interested in working on.

*Pick your characters.* Definitely include yourself. Include other people you care about as well as people you think are relevant to the problem(s) you're working with. If you're feeling adventurous, include people who don't believe you yet.

**Step 3.** *Dream.*

*Write a story about what your place/time is like after solving the problem(s) your group decided on.* This is best done individually and in a quiet room. Laptops or pen and paper are equally fine. Word count is less important than expansive imagination. Take at least thirty minutes to do this and know that it won't be enough time to imagine a future Earth. (You could write for two hours and it still won't be enough.) Story ideas are a conversation between two people in the future or a letter from a future person to a present-day person (such as me in 2060 writing to my nieces, nephews, and other siblings).

**Step 4.** *Discuss.*

Share with one another what you envisioned. You can swap stories, read stories out loud, or just discuss what you each wrote.

Pay attention to similarities and differences in your visions. Do you notice any patterns?

**Step 5.** *Next steps.*

Decide what you want to do next. You could:

- Do a second round of writing, adding to the stories you wrote and incorporating elements (characters, solutions, aspects of setting) from the stories of other group members. Share again and watch magic unfurl!
- Share the stories with people outside your group. Pick a deadline, polish up what you produced, and figure out how you want them to be shared. You could:
  - —put them together to be shared in paper form, like a zine or small booklet.
  - —put them online on a blog or someone's website.
  - —email them to your friends as inspiration for them to come to the next visioning session.

  Whatever you decide, it's most important that you share your vision(s) with people in the community you dreamed about. That's where you'll have the most influence and impact. You can also share them at www.ericholthaus.com/futureearthstories.
- Host another gathering with new, different, or more people. Don't forget to update your Why, or make sure your original reasoning still sticks.
- Do anything else your heart desires.

## REFLECTION QUESTIONS

- What are the identities of people in your group? What are their genders, races, and classes? Do they all speak the same language and with the same ability? What is their relationship to children? What is their citizenship status? What identities are present in your neighborhood that aren't present in your group? How might you invite people with those identities next time? How might you support them to have their own visioning activity in preparation for swapping stories, insights, and ideas in the future?
- What elements of the present/past created or shaped the future you imagined? Discuss with your group. If you aren't sure, research these elements before the next gathering.
- Discuss what is preventing the worlds you imagined from becoming a reality. Are there local ways your group could help remove those barriers? If you're already working on it, are there ways you could connect with groups in other places working to remove the same barriers?

Most of all, have fun and don't hold back. Remember, you are building a world that is going to be irresistible.

# ACKNOWLEDGMENTS

This book is my effort to imagine a radical vision for what our future could look like. These pages are my love letter to the world, and like every love letter, sending it causes me to feel very vulnerable, but I'm ready.

Most of the work on this project happened on the occupied homelands of the Mdewakanton Dakota people, but also on the lands of the Klallam, Tohono O'odham, Kaw, Lenape, and Haudenosaunee Confederacy. For hundreds of years the Mdewakanton Dakota people lived in harmony with the world around them, especially the waters of the Mnisóta Wakpá, which gave them life. When European settlers arrived, they were violently forced off their lands and their families were broken up, in an attempt to erase them. But they are still here, and we have a lot to learn from them.

Acknowledging this history is not enough. I know I must work to repair the historical damage that I personally benefit from, and work toward a world that centers on justice. That spirit of repair guided this entire project.

In my consultation with Indigenous peoples as part of this project, especially with peoples from the Marshall Islands, I have learned that the future is never fated. We are always reinventing ourselves. That's part of what makes us human. The climate emergency is not the first existential threat we have faced as a species, and it's possible for us to radically change course. But to do that we must listen to people with radically different experiences and views.

I've also attempted to do this project as a low-carbon-reporting experiment. Instead of personally traveling to the key places I describe, I've tried to let people from these places tell their own stories as much as possible, through both phone calls and writings. I started this project fascinated by the Marshall Islands, and I tried to speak with as many people there as I could. I then learned the stories of people across Micronesia, about how their lives were bound up not only by the pollution from distant countries but also by the nuclear legacy and the economic pressures of colonialism. In the five years it took to compile this book, I spoke with hundreds of people in dozens of countries. I didn't talk with as many people as I wanted to, but this is just the start of a reporting philosophy I will be continuing for the rest of my life.

In preparing this project, I'm sure I've made mistakes and will continue to do so. I apologize for any further harm I have caused in mischaracterizing the challenges we face from my vantage point. Putting this book together has been an exploration of my own privilege. It's especially important for me to make sure that I'm doing this in a way that is respectful. I've learned that I've still got a lot to learn—but the hopeful part is that the futures that are still possible will be unlike anything I've ever imagined. This book has changed me, and I hope it changes you too.

This work would not have been possible without the tireless support of my agent, Brandi Bowles, and my editor, Miles Doyle, whose patience is truly legendary. Thanks also to my primary day-job editors over the past five years: Eliza Anyangwe, Rob Wijnberg, Nikhil Swaminathan, Darby Minow Smith, and Torie Bosch—especially Torie, who basically taught me how to write. And special thanks to Caroline Contillo and Lawrence Barriner II, who crafted the grief and imagination exercises. My hope is that they will help spark a movement that will change the world.

Some of the ideas and writings included in this book I previously published in *Rolling Stone*, *Grist*, *Slate*, *Pacific Standard* (RIP), *Quartz*, *ThinkProgress* (RIP), and *The Correspondent*. I am grateful for the opportunities I've had in these and other publications to contribute my work publicly over the years. Freelance journalists deserve your support. Support them.

I'm especially grateful for the moral support from my parents, Les and Janet Holthaus, my sister, Jennifer Atchison, and Karen Edquist, Karin Aebersold, Emily Sharp, Katy Backes Kozhimannil, Patrick Schmitt, and everyone else who let me talk for untold hours about how stressed I was, and how I'd never finish. I finished, and it happened because of you. Thanks to everyone who sent their children's drawings of ladybugs and anglerfish to decorate our home. Thanks to the person or persons who compiled the "Deep Focus" playlist on Spotify, and all the ASMR and EDM artists who kept me calm and helped me concentrate. Thanks to the staff at the Seward Cafe and at Groundswell, where a lot of this writing took place. Thanks to all the farmers who grew the coffee beans that kept me going. Thanks also to my Patreon supporters who helped keep our family afloat during the ups and downs of this project.

My evolution as a writer came out of necessity. My training was in meteorology, but I've always been drawn to understanding the social justice of weather. Finding the language of moral clarity and justice has been ongoing, but these people, and so many more whom I haven't included, have especially shaped my path: Kathy Jentil-Kijiner, Paul Robbins, Mark Chmiel, Katherine Crocker, Sarah Myhre, Sydney Ghazarian, Justine Calma, Kaitlin Curtice, Winston Hearn, Tanmoy Goswami, Irene Caselli, OluTimehin Adegbeye, and Nesrine Malik.

I am very grateful for clear storytelling on climate and countless conversations over the years with Jennie Ferrara, Jamie Margolin, Julia Steinberger, Nathan Thanki, Renee Lertzman, Faith Kearns, Alex Steffen, Mary Annaïse Heglar, Jacquelyn Gill, Kevin Anderson, Glen Peters, Karthik Ganapathy, Pablo Suarez, Marjorie Brans, Mengesha Gebremichel, Michael Norton, Jon Foley, Deke Arndt, Dargan Frierson, Brian Kahn, Kate Knuth, Maria Langholz, and so many others, on and off climate Twitter.

I forget now who gave me the best advice I got throughout this process: "The key to writing a good book is to write a bad book, and then fix it." That goes for everything we do, every day.

The only things that matter are that we struggle together, that we learn from one another, and that our work bends toward justice.

Thanks especially to Roscoe and Zeke, for holding my hand and taking so long to go to bed, and for giving me the courage to imagine a better world. This book is dedicated to you.

*Eric Holthaus*

*November 2019*

# NOTES

## A Living Emergency

3 With no other options: Rachel Becker, “EPA Says Puerto Rico Residents Resorted to Contaminated Water at Dorado Superfund,” *The Verge*, October 13, 2017, www.theverge.com/2017/10/13/16474428/puerto-rico-hurricane-maria-superfund-site-water-shortage-epa-recovery.

4 “[Today’s] the first time I saw”: Personal communication via text message, September 26, 2017.

4 A 2019 study in the journal: David Keellings and José J. Hernández Ayala, “Extreme Rainfall Associated with Hurricane Maria over Puerto Rico and Its Connections to Climate Variability and Change,” *Geophysical Research Letters* 46, no. 5 (March 2019): 2964–73, https://agupubs.onlinelibrary.wiley.com/doi/10.1029/2019GL082077.

4 Lead author David Keellings told: American Geophysical Union, “Climate Change to Blame for Hurricane Maria’s Extreme Rainfall,” news release, April 16, 2019, https://phys.org/news/2019-04-climate-blame-hurricane-maria-extreme.html.

4 Hurricane Maria damaged or destroyed: Earth Institute at Columbia University, “Hurricane Maria Study Warns: Future Climate-Driven Storms May Raze Many Tropical Forests,” news release, March 25, 2019, www.eurekalert.org/pub_releases/2019-03/eiac-hms032119.php.

5 “the largest psychosocial disaster in the United States”: Justine Calma, “The Storm Isn’t Over,” *Grist*, January 8, 2018, https://grist.org/briefly/hurricane-survivors-are-still-dealing-with-the-emotional-toll-of-2017s-horrific-storms/.

7 A 2018 study found: Blaine Friedlander, "Severe Caribbean Droughts May Magnify Food Insecurity," Cornell Chronicle, November 5, 2018, news.cornell.edu/stories/2018/11/severe-caribbean-droughts-may-magnify-food-insecurity.

7 Fijian president Jioji Konrote vowed: Kate Wheeling, "After a Record-Breaking Atlantic Hurricane Season, Here's What to Expect in the South Pacific," *Pacific Standard*, November 3, 2017, https://psmag.com/environment/cop23-what-to-expect-from-cyclone-season-in-south-pacific.

8 the UN called it: "Cyclone Idai: Emergency Getting 'Bigger by the Hour,' Warns UN Food Agency," UN News, March 19, 2019, https://news.un.org/en/story/2019/03/1034951.

9 The permafrost—frozen soil: Louise M. Farquharson, Vladimir E. Romanovsky, William L. Cable, Donald A. Walker, Steven V. Kokelj, and Dmitry Nicolsky, "Climate Change Drives Widespread and Rapid Thermokarst Development in Very Cold Permafrost in the Canadian High Arctic," *Geophysical Research Letters* 46, no. 12 (June 2019): 6681–89, https://agupubs.onlinelibrary.wiley.com/doi/full/10.1029/2019GL082187.

9 first time in tens of thousands of years: Samson Reiny, "Arctic Shifts to a Carbon Source due to Winter Soil Emissions," NASA, November 8, 2019, https://www.nasa.gov/feature/goddard/2019/arctic-shifts-to-a-carbon-source-due-to-winter-soil-emissions.

9 July 2019 was the hottest month: Mario Picazo, "July 2019 Officially Hottest Month Ever Measured on Earth," The Weather Network, August 6, 2019, https://www.theweathernetwork.com/ca/news/article/july-2019-officially-hottest-month-ever-measured-on-earth.

10 "Every morning, you wake up": Jacqueline Charles, "Why Some Hurricane Dorian Survivors Are Staying on Abaco: 'We Believe in Bouncing Back,'" *Miami Herald*, September 14, 2019, https://www.miamiherald.com/news/weather/hurricane/article235043772.html.

10 rapidly advancing flames on all sides: "Bushfires Crisis: Thousands Stranded in New Year's Eve Fury," *The Australian*, n.d., https://www.theaustralian.com.au/nation/bushfires-crisis-thousands-stranded-in-new-years-eve-fury/news-story/b65fc8b80c92fa31c79cfbd91c12aa44.

10 480 million mammals, birds, and reptiles were killed: "A Statement About the 480 Million Animals Killed in NSW Bushfires Since September," University of Sydney, January 3, 2020, https://sydney.edu.au/news-opinion/news/2020/01/03/a-statement-about-the-480-million-animals-killed-in-nsw-bushfire.html.

10 Alaska and Greenland: Alan J. Parkinson et al., "Climate Change and Infectious Diseases in the Arctic: Establishment of a Circumpolar Working Group," *International Journal of Circumpolar Health*, September 30, 2014, https://www.ncbi.nlm.nih.gov/pmc/articles/PMC4185088/.

11 Heat waves have become: Ethan D. Coffel, Radley M. Horton, and Alex de Sherbinin, "Temperature and Humidity Based Projections of a Rapid Rise in Global Heat Stress Exposure During the 21st Century," *Environmental Research Letters* 13, no. 1 (December 2017): 014001, https://iopscience.iop.org/article/10.1088/1748-9326/aaa00e/meta.

11 Young people growing up today: Sophie Bethune, "Gen Z More Likely to Report Mental Health Concerns," *American Psychological Association* 50, no. 1 (January 2019), https://www.apa.org/monitor/2019/01/gen-z.

11 In recent years, about a quarter: Bruce C. Forbes, Timo Kumpula, Nina Meschtyb, Roza Laptander, Marc Macias-Fauria, Pentii Zetterberg, Mariana Verdonen, et al., "Sea Ice, Rain-on-Snow and Tundra Reindeer Nomadism in Arctic Russia," *Biology Letters* 12, no. 11 (November 2016): 20160466, https://royalsocietypublishing.org/doi/full/10.1098/rsbl.2016.0466.

12 "Postcards from the Anthropocene": Brian Kahn, Twitter post, June 6, 2018, https://twitter.com/blkahn/status/1004404764766023680.

13 By the 2040s: Union of Concerned Scientists, *Underwater: Rising Seas, Chronic Floods, and the Implications for US Coastal Real Estate*, report, June 18, 2018, https://www.ucsusa.org/resources/underwater.

13 "the loss of all coastal cities": James Hansen, "Ice Melt, Sea Level Rise and Superstorms: The Threat of Irreparable Harm," March 22, 2016, http://www.columbia.edu/~jeh1/mailings/2016/20160322_IrreparableHarm.pdf.

14 “Imagine a world where”: “Jellyfish Are Taking Over!” *Living on Earth*, December 2, 2016, https://loe.org/shows/segments.html?programID=16-P13-00049&segmentID=5.

15 Using records of everything: Cynthia Rosenzweig, David Karoly, Marta Vicarelli, Peter Neofotis, Qigang Wu, Gino Casassa, Annette Menzel, et al., “Attributing Physical and Biological Impacts to Anthropogenic Climate Change,” *Nature* 453 (May 2008): 353–57, accessed at Climate Signals website, https://www.climatesignals.org/scientific-reports/attributing-physical-and-biological-impacts-anthropogenic-climate-change.

16 “If we don’t take urgent action”: Personal communication via email, February 19, 2019.

16 can lapse into solastalgia: Glenn Albrecht, “The Age of Solastalgia,” The Conversation, August 7, 2012, https://theconversation.com/the-age-of-solastalgia-8337.

16 two- to eight-year record: Frances C. Moore, Nick Obradovich, Flavio Lehner, and Patrick Baylis, “Rapidly Declining Remarkability of Temperature Anomalies May Obscure Public Perception of Climate Change,” *Proceedings of the National Academy of Sciences of the United States of America* 116, no. 11 (March 12, 2019), https://www.pnas.org/content/116/11/4905.

18 may have already started: Sarah Fecht, “Current Megadrought in the West Could Be One of the Worst in History,” *State of the Planet*, Earth Institute, Columbia University, December 13, 2018, https://blogs.ei.columbia.edu/2018/12/13/megadrought-west-climate-change/.

20 about love: Sarah Myhre, Twitter post, https://twitter.com/SarahEMyhre/status/1164211181013196801?s=03).

21 “If we don’t demand radical change”: Naomi Klein, “Let Them Drown: The Violence of Othering in a Warming World,” *London Review of Books* 38, no. 11 (June 2, 2016), https://www.lrb.co.uk/v38/n11/naomi-klein/let-them-drown.

24 In the waning days of 2018: Intergovernmental Panel on Climate Change (IPCC), *Special Report: Global Warming of 1.5°C*, 2018, https://www.ipcc.ch/sr15/.

25 “If action is not taken”: Question posed during IPCC press conference, October 8, 2018, via weblink, starting at 47:58, https://www.youtube.com/watch?v=12S3dKrxj7c.

25 A team of scientists: Intergovernmental Science-Policy Platform on Biodiversity and Ecosystem Services (IPBES), "Nature's Dangerous Decline 'Unprecedented'; Species Extinction Rates 'Accelerating,'" news release, May 4, 2019, https://ipbes.net/news/Media-Release-Global-Assessment.

25 "Human Society Under Urgent Threat": Jonathan Watts, "Human Society Under Urgent Threat from Loss of Earth's Natural Life," *The Guardian*, May 6, 2019, https://www.theguardian.com/environment/2019/may/06/human-society-under-urgent-threat-loss-earth-natural-life-un-report.

26 What we're doing to the planet's climate: Melissa Davey, "Humans Causing Climate to Change 170 Times Faster than Natural Forces," *The Guardian*, February 12, 2017, https://www.theguardian.com/environment/2017/feb/12/humans-causing-climate-to-change-170-times-faster-than-natural-forces.

26 "the human magnitude of climate change": Melissa Davey, "Humans Are Changing Climate Faster Than Natural Forces," Climate Central, February 19, 2017, https://www.climatecentral.org/news/humans-changing-climate-faster-than-natural-forces-21175.

26 "The old ways of conservation": Personal communication via phone, January 16, 2019.

27 "This is a huge": Michael Innis, "Climate-Related Death of Coral Around World Alarms Scientists," *New York Times*, April 9, 2016, https://www.nytimes.com/2016/04/10/world/asia/climate-related-death-of-coral-around-world-alarms-scientists.html.

28 "It's a wake-up call": Personal communication via phone, April 9, 2016.

28 "This is a story that the rest": Eric Holthaus, "The Largest Coral Atoll in the World Lost 80 Percent of Its Coral to Bleaching," *ThinkProgress*, April 12, 2016, https://thinkprogress.org/the-largest-coral-atoll-in-the-world-lost-80-percent-of-its-coral-to-bleaching-168ba23b0562/.

29 "The World Has Just over": Chris Mooney and Brady Dennis, "The World Has Just over a Decade to Get Climate Change Under Control, U.N. Scientists Say," *Washington Post*, October 7, 2018, https://www.washingtonpost.com/energy-environment/2018/10/08/world-has-only-years-get-climate-change-under-control-un-scientists-say/.

29 "rapid, far-reaching": *Special Report: Global Warming of 1.5°C*, Intergovernmental Panel on Climate Change, 2018, https://www.ipcc.ch/sr15/chapter/spm/.

30 "All options need to be exercised": IPCC press conference, October 8, 2018, https://www.youtube.com/watch?v=12S3dKrxj7c.

30 "Sometimes, when I get out of bed": Personal communication via phone, May 9, 2019.

31 "One of the very few benefits": Personal communication via phone, May 9, 2019.

32 "We're at that time where": Personal communication via phone, February 26, 2019.

33 "The major problem with society": Personal communication via phone, February 26, 2019.

33 "Liminal space is a time of radical uncertainty": Personal communication via phone, February 26, 2019.

34 "Climate change is first and foremost": Personal communication via phone, November 29, 2016.

34 A recent study mapped out: Stockholm Resilience Centre, "Curbing Emissions with a New 'Carbon Law,'" Stockholm University, March 23, 2017, https://www.stockholmresilience.org/research/research-news/2017-03-23-curbing-emissions-with-a-new-carbon-law.html.

34 "It's way more than adding": Brad Plumer, "Scientists Made a Detailed 'Roadmap' for Meeting the Paris Goals. It's Eye-Opening," *Vox*, March 24, 2017, https://www.vox.com/energy-and-environment/2017/3/23/15028480/roadmap-paris-climate-goals.

36 A recent survey revealed: Elena Berton, "Flight Shaming Hits Air Travel as 'Greta Effect' Takes Off," Reuters, October 2, 2019, https://www.reuters.com/article/us-travel-flying-climate/flight-shaming-hits-air-travel-as-greta-effect-takes-off-idUSKBN1WH23G.

37 "Our civilization is being sacrificed": "Greta Thunberg Full Speech at UN Climate Change COP24 Conference," December 15, 2018, https://www.youtube.com/watch?v=VFkQSGyeCWg.

39 "I am not saying anything new": Greta Thunberg, Facebook post, February 11, 2019, https://www.facebook.com/gretathunbergsweden/photos/a.733630957004727/773673599667129/?type=1&theater.

42 "dangerous anthropogenic interference": United Nations, *Report of the Intergovernmental Negotiating Committee for a Framework Convention on Climate Change on the Work of the Second Part of Its Fifth Session, Held at New York from 30 April to 9 May 1992*, https://unfccc.int/sites/default/files/resource/docs/a/18p2a01.pdf.

43 there is no "we" that is causing climate change: Genevieve Guenther, "Who Is the *We* in 'We Are Causing Climate Change'?," *Slate*, October 10, 2018, https://slate.com/technology/2018/10/who-is-we-causing-climate-change.html.

43 fill you with rage: Amy Westervelt, "The Case for Climate Rage," Popula, August 19, 2019, https://popula.com/2019/08/19/the-case-for-climate-rage/.

45 In a 2018 essay for On Being: Kate Marvel, "We Need Courage, Not Hope, to Face Climate Change," On Being blog, March 1, 2018, https://onbeing.org/blog/kate-marvel-we-need-courage-not-hope-to-face-climate-change/.

47 "There has been more openness": Personal communication via phone, February 15, 2019.

48 "The decisions we make now": Matt McGrath, "'Reasons to Be Hopeful' on 1.5C Global Temperature Target," BBC News, October 3, 2018, https://www.bbc.com/news/science-environment-45720740.

48 "We're talking about the kind of crisis": Marlow Hood, "UN Report on 'Mission Impossible' Climate Target: Key Points," Phys.org, October 1, 2018, https://phys.org/news/2018-10-mission-impossible-climate-key.html.

51 "From my point of view": Personal communication via phone, February 26, 2019.

53 Indigenous peoples in the Americas: Julian Brave Noisecat, "We Need Indigenous Wisdom to Survive the Apocalypse," *The Walrus*, October 18, 2019, https://thewalrus.ca/we-need-indigenous-wisdom-to-survive-the-apocalypse/.

53 "Look at what colonialism": Personal communication via phone, February 20, 2019.

53 "I don't think that we're ever": Personal communication via phone, February 20, 2019.

53 is rooted in international law: "In Shadow of #MeToo: The Coming Reckoning on Consent and Climate Change," Fawn Sharp and Matthew Randazzo V, Crosscut, April 30, 2019, https://crosscut.com/2019/04/shadow-metoo-coming-reckoning-consent-and-climate-change.

54 "Indigenous people are already": Personal communication via phone, February 20, 2019.

54 "Improving people's behavior": Personal communication via phone, February 20, 2019.

55 "When we finally have people": Personal communication via phone, February 26, 2019.

55 "As for climate change": Personal communication via phone, February 26, 2019.

57 "the very ability to begin bringing": Personal communication via phone, January 9, 2019.

59 "Our extractive, wasteful": Alexandria Ocasio-Cortez, Twitter post, May 6, 2019, https://twitter.com/AOC/status/1125432683448950784.

60 "We're not just": Personal communication via phone, January 18, 2019.

61 "civilization could crumble in our lifetime": Sunrise Movement town hall meeting at Macalester College, April 25, 2019, https://www.facebook.com/sunriseminnesota/posts/this-is-a-live-stream-of-macalesters-sunrise-town-hall-meeting-for-a-green-new-d/2223756721049255/.

62 "We will destroy out of love": Personal notes at Sunrise Movement town hall meeting at Macalester College, April 25, 2019, https://www.facebook.com/sunriseminnesota/posts/this-is-a-live-stream-of-macalesters-sunrise-town-hall-meeting-for-a-green-new-d/2223756721049255/.

62 "Your story can change the debate": Sunrise Movement town hall meeting at Macalester College, April 25, 2019.

63 "There are people calling for": Personal communication via phone, January 18, 2019.

64 "a new national, social": 116th Congress, "H.Res.109 Recognizing the Duty of the Federal Government to Create a Green New

Deal," introduced February 7, 2019, https://www.congress.gov/bill/116th-congress/house-resolution/109/text.

64 "I think that this is a very special": Danielle Kurtzleben, "Rep. Alexandria Ocasio-Cortez Releases Green New Deal Outline," February 7, 2019, NPR, https://www.npr.org/2019/02/07/691997301/rep-alexandria-ocasio-cortez-releases-green-new-deal-outline.

65 In September 2019: "Thousands in New Zealand Kick-Start New Wave of Climate Protests," *Aljazeera*, September 26, 2019, https://www.aljazeera.com/news/2019/09/thousands-zealand-kickstart-wave-climate-protests-190927033920880.html.

65 the country's prime minister signed: Zoe Tidman, "New Zealand Passes 'Zero Carbon' Law in Fight Against Climate Change," *Independent*, November 7, 2019, https://www.independent.co.uk/news/world/australasia/new-zealand-zero-carbon-law-emissions-jacinda-ardern-climate-change-a9189341.html.

67 IPCC's "best-case" scenario: "Impacts of 1.5°C of Global Warming on Natural and Human Systems," chap. 3, http://iacweb.ethz.ch/staff/sonia/download/etc/IPCC_SR15_Storylines/IPCC_SR15_Storylines.pdf.

68 "is no longer recognizable": "Impacts of 1.5°C of Global Warming on Natural and Human Systems," chap. 3, http://iacweb.ethz.ch/staff/sonia/download/etc/IPCC_SR15_Storylines/IPCC_SR15_Storylines.pdf.

68 "Droughts and stress": "Impacts of 1.5°C of Global Warming on Natural and Human Systems," chap. 3, http://iacweb.ethz.ch/staff/sonia/download/etc/IPCC_SR15_Storylines/IPCC_SR15_Storylines.pdf.

**2020–2030: Catastrophic Success**

73 "I took a video": Junior, "Yangdidi: Stories from Super Typhoon Maysak Survivors," December 24, 2015, http://blogs.ubc.ca/yangdidi/2015/12/24/junior/.

75 Reporting for the Solutions Journalism Network: "Typhoon Haiyan Forces an Entire Island Community to Relocate," PRI, November 22, 2013, https://www.pri.org/stories/2013-11-22/typhoon-haiyan-forces-entire-island-community-relocate.

76 "a wasteland of mud and debris": "Typhoon Haiyan: Philippines Battles to Bring Storm Aid," BBC News, November 10, 2013, https://www.bbc.com/news/world-asia-24887746.

76 "We may have ratified our own doom": "Typhoon Haiyan," https://www.bbc.com/news/world-asia-24887746.

78 "I kept thinking the whole world": Personal communication via phone, November 29, 2016.

79 "high ambition coalition": Ed King, "Tony de Brum: The Emerging Climate Champion at COP21," Climate Home News, October 12, 2015, https://www.climatechangenews.com/2015/12/10/tony-de-brum-the-emerging-climate-champion-at-cop21/.

79 "small island girl with big dreams": Selina Leem, "Marshall Islands 18-Year-Old Thanks UN for Climate Pact," Climate Change News, December 14, 2015, https://www.climatechangenews.com/2015/12/14/marshall-islands-18-year-old-thanks-un-for-climate-pact/.

80 "If we do have to lose our islands": Personal communication via phone, November 29, 2016.

81 "we just can't afford it": Personal communication via phone, November 29, 2016.

81 "It just hit me": Personal communication via phone, November 29, 2016.

82 "This agreement is for": Leem, "Marshall Islands 18-Year-Old Thanks UN for Climate Pact," https://www.climatechangenews.com/2015/12/14/marshall-islands-18-year-old-thanks-un-for-climate-pact/.

82 Multiple studies have now shown: Guy J. Abel, Michael Brottrager, Jesus Crespo Cuaresma, and Raya Muttarak, "Climate, Conflict, and Forced Migration," *Global Environmental Change* 54 (January 2019): 239–49, https://www.sciencedirect.com/science/article/pii/S0959378018301596?via%3Dihub.

83 The Pentagon has warned: Caitlin Werrell and Francesco Femia, "Climate Change and National Security in the 2014 Quadrennial Defense Review," Center for Climate and Security, March 4, 2014, https://climateandsecurity.org/2014/03/04/climate-change-and-national-security-in-the-2014-quadrennial-defense-review/.

83 "Displacement of populations": "Marshalls Likens Climate Change Migration to Cultural Genocide," RNZ, October 6, 2015, https://www.rnz.co.nz/international/pacific-news/286139/marshalls-likens-climate-change-migration-to-cultural-genocide.

83 "In 10 years we drown": Ted Scheinman, "COP22 in Review: We'll Work Until We Drown," *Pacific Standard*, November 18, 2016, https://psmag.com/news/cop22-in-review-well-work-until-we-drown.

85 "shall determine the existence": "Charter of the United Nations, Chapter VII: Action with Respect to Threats to the Peace, Breaches of the Peace, and Acts of Aggression," United Nations, https://www.un.org/en/sections/un-charter/chapter-vii/.

85 "This effort could also spark": Personal communication via phone, February 1, 2017.

86 "a wicked problem": Personal communication via phone, February 1, 2017.

87 "I think the countries of the world": Personal communication via phone, February 1, 2017.

88 "International law recognizes": Michael B. Gerrard, "America Is the Worst Polluter in the History of the World. We Should Let Climate Change Refugees Resettle Here," *Washington Post*, June 25, 2015, https://www.washingtonpost.com/opinions/america-is-the-worst-polluter-in-the-history-of-the-world-we-should-let-climate-change-refugees-resettle-here/2015/06/25/28a55238-1a9c-11e5-ab92-c75ae6ab94b5_story.html.

88 "is radical, rapid reductions": Personal communication via phone, February 1, 2017.

89 "I get a lot of people": Personal communication via phone, December 1, 2016.

92 "goodbye": Deke Arndt, Twitter post, June 5, 2019, https://twitter.com/DekeArndt/status/1136259382151245825.

95 "eco-fascism": Jason Wilson, "Eco-fascism Is Undergoing a Revival in the Fetid Culture of the Extreme Right, *The Guardian*, March 19, 2019, https://www.theguardian.com/world/commentisfree/2019/mar/20/eco-fascism-is-undergoing-a-revival-in-the-fetid-culture-of-the-extreme-right.

95 A group of scientists in Hawaii: Camilo Mora, Daniele Spirandelli, Erik C. Franklin, John Lynham, Michael B. Kantar, Wendy Miles, Charlotte Z. Smith, et al., "Broad Threat to Humanity from Cumulative Climate Hazards Intensified by Greenhouse Gas Emissions," *Nature Climate Change* 8 (December 2018): 1062–71, https://doi.org/10.1038/s41558-018-0315-6.

95 "None of this happens in a vacuum": Eric Holthaus, "'Climate Change War' Is Not a Metaphor," *Slate*, April 18, 2014, https://slate.com/technology/2014/04/david-titley-climate-change-war-an-interview-with-the-retired-rear-admiral-of-the-navy.html.

96 divestment movement began to snowball: Rachel Koning Beals, "Goldman Sachs Becomes First Major U.S. Bank to Stop Funding Arctic Drilling, Pulls Back on Coal," MarketWatch, December 21, 2019, https://www.marketwatch.com/story/goldman-sachs-becomes-first-major-us-bank-to-stop-funding-arctic-drilling-pulls-back-on-coal-2019-12-16?link=sfmw_tw.

101 "We lost our home": Personal communication via Twitter message, October 18, 2018.

101 "You [started] to wonder how long": Personal communication via phone, January 12, 2019.

103 "Prepare for catastrophic success": Holthaus, "'Climate Change War' Is Not a Metaphor," https://slate.com/technology/2014/04/david-titley-climate-change-war-an-interview-with-the-retired-rear-admiral-of-the-navy.html.

104 "Because we built such": Personal communication via phone, January 18, 2019.

107 Bernie Sanders's Green New Deal: "The Green New Deal," Bernie Sanders.com, accessed January 15, 2020, https://berniesanders.com/en/issues/green-new-deal/.

107 by 2030: "The Global Price Tag for 100 Percent Renewable Energy: $73 Trillion," Yale Environment 360, December 20, 2019, https://e360.yale.edu/digest/the-global-price-tag-for-100-percent-renewable-energy-73-trillion.

108 "The principles of right to self-determination": Personal communication via phone, February 19, 2019.

110 "The issue throughout the Caribbean": Personal communication via phone, February 15, 2019.

113 Netherlands Supreme Court ruled: Jelmer Mommers, "Thanks to This Landmark Court Ruling, Climate Action Is Now Inseparable from Human Rights," *The Correspondent*, December 2019, https://thecorrespondent.com/194/thanks-to-this-landmark-court-ruling-climate-action-is-now-inseparable-from-human-rights/25683007966-93bb1751.

113 pathways to ambitious climate action: Jelmer Mommers, "Lawyers Are Going to Court to Stop Climate Change. And It Might Just Work," *The Correspondent*, December 2019, https://thecorrespondent.com/185/lawyers-are-going-to-court-to-stop-climate-change-and-it-might-just-work/24491528215-2ac4e218.

114 filed a lawsuit in 2015: Eric Holthaus, "Children Sue Over Climate Change," *Slate*, November 16, 2015, https://slate.com/news-and politics/2015/11/children-sue-the-obama-administration-over-climate-change.html.

114 "In these proceedings": *Juliana v. United States*, United States Court of Appeals for Ninth Circuit, http://blogs2.law.columbia.edu/climate-change-litigation/wp-content/uploads/sites/16/case-documents/2020/20200117_docket-18-36082_opinion.pdf.

115 "The idea that the International Criminal Court": Personal communication via phone, February 8, 2019.

116 legal personhood status: Ananya Bhattacharya, "Birds to Holy Rivers: A List of Everything India Considers 'Legal Persons,'" Quartz India, June 7, 2019, https://qz.com/india/1636326/who-apart-from-human-beings-are-legal-persons-in-india/amp/?__twitter_impression=true.

116 In the Cook Islands: Celeste Coughlin, "Rights of the Pacific Ocean Initiative," Earth Law Center, April 23, 2019, https://www.earthlawcenter.org/blog-entries/2019/4/rights-of-the-pacific-ocean-initiative.

**2030–2040: Radical Stewardship**

123 "In a circular, or cyclical, economy": Personal communication via phone, March 1, 2019.

123 repair and maintenance: Shannon Mattern, "Maintenance and Care," *Places Journal*, November 2018, https://placesjournal.org/article/maintenance-and-care/?cn-reloaded=1&cn-reloaded=1.

123 "Today's linear and degenerative": Personal communication via phone, March 1, 2019.

125 "let's just rethink thriving": Personal communication via phone, March 1, 2019.

126 "By 2040, we started to move back": Personal communication via phone, March 1, 2019.

128 "There's an unprecedented opportunity": Personal communication via phone, March 1, 2019.

133 "In the aftermath of the storm": Personal communication via phone, February 15, 2019.

137 "We're afraid equity means": Peter Callaghan, "In First Speech as Mayor, Melvin Carter Celebrates—and Challenges—Residents of St. Paul," *MinnPost*, January 3, 2018, https://www.minnpost.com/politics-policy/2018/01/first-speech-mayor-melvin-carter-celebrates-and-challenges-residents-st-paul/.

139 "We all exist in a long line": Callaghan, "In First Speech as Mayor, Melvin Carter Celebrates—and Challenges—Residents of St. Paul," https://www.minnpost.com/politics-policy/2018/01/first-speech-mayor-melvin-carter-celebrates-and-challenges-residents-st-paul/.

140 "Building in a forest": Personal communication via phone, January 12, 2019.

142 One study found that using wood: Andrius Bialyj, "The Future of Architecture: CLT Wooden Skyscrapers," AGA CAD website blog, July 5, 2019, www.aga-cad.com/blog/the-future-of-architecture-clt-wooden-skyscrapers.

150 "We would see changes in land use": Personal communication via phone, March 1, 2019.

155 water from the Ems: "In Face of Rising Sea Levels the Netherlands 'Must Consider Controlled Withdrawal," Vrij Nederland, February 9, 2019, https://www.vn.nl/rising-sea-levels-netherlands/.

156 Instead of a three-foot increase: Robert M. DeConto and David Pollard, "Contribution of Antarctica to Past and Future Sea-Level Rise," *Nature* 531 (March 2016): 591–97, https://www.nature.com/articles/nature17145.

156 in line with predictions: Jeff Tollefson, "Antarctic Model Raises Prospect of Unstoppable Ice Collapse," *Nature*, March 30, 2016, https://www.nature.com/news/antarctic-model-raises-prospect-of-unstoppable-ice-collapse-1.19638.

157 "Every revision to our understanding": Personal communication via phone, November 2, 2017.

157 "We didn't predict that Pine Island": Eric Holthaus, "Ice Apocalypse," Grist, November 21, 2017, https://grist.org/article/antarctica-doomsday-glaciers-could-flood-coastal-cities/.

## 2040–2050: New Technologies and New Spiritualities

160 "Geoengineering holds forth": Newt Gingrich, "Stop the Green Pig: Defeat the Boxer-Warner-Lieberman Green Pork Bill Capping American Jobs and Trading America's Future, *Human Events*, June 3, 2008, https://humanevents.com/2008/06/03/stop-the-green-pig-defeat-the-boxerwarnerlieberman-green-pork-bill-capping-american-jobs-and-trading-americas-future/.

161 Take them away: Chelsea Harvey, "Cleaning Up Air Pollution May Strengthen Global Warming," *Scientific American*, January 22, 2018, https://www.scientificamerican.com/article/cleaning-up-air-pollution-may-strengthen-global-warming/.

162 "Thunderstorms in China": Zin Yang et al., "Distinct Weekly Cycles of Thunderstorms and Potential Connection with Aerosol Type in China," *Geophysical Research Letters* 43, no. 16 (August 28, 2016), https://agupubs.onlinelibrary.wiley.com/doi/full/10.1002/2016GL070375.

163 Research in 2019 showed: Drew Shindell and Christopher J. Smith, "Climate and Air-Quality Benefits of a Realistic Phase-Out of Fossil Fuels," *Nature* 573 (September 19, 2019), https://doi.org/10.1038/s41586-019-1554-z.

163 "This is known territory": Eric Holthaus, "Devil's Bargain," Grist, February 8, 2018, https://grist.org/article/geoengineering-climate-change-air-pollution-save-planet/.

163 One recent study estimated: Kevin Loria, "A Last-Resort 'Planet-Hacking' Plan Could Make Earth Habitable for Longer—But Scientists Warn It Could Have Dramatic Consequences," *Business Insider*, July 20, 2017, https://www.businessinsider.com/geoengineering-technology-could-cool-the-planet-2017-7.

164 "It's very plausible that": Holthaus, "Devil's Bargain," https://grist.org/article/geoengineering-climate-change-air-pollution-save-planet/.

164 "Geoengineering is like taking painkillers": Leah Burrows, "Mitigating the Risk of Geoengineering," *The Harvard Gazette*, December 12, 2016, https://news.harvard.edu/gazette/story/2016/12/mitigating-the-risk-of-geoengineering/.

165 One of the biggest risks: Christopher H. Trisos et al., "Potentially Dangerous Consequences for Biodiversity of Solar Geoengineering Implementation and Termination," *Nature Ecology and Evolution* 2 (2018): 475–82, https://www.nature.com/articles/s41559-017-0431-0.

165 "I could imagine global conflicts": Holthaus, "Devil's Bargain," https://grist.org/article/geoengineering-climate-change-air-pollution-save-planet/.

165 A study from researchers: Solomon M. Hsiang et al., "Quantifying the Influence of Climate on Human Conflict," *Science* 341, no. 6151 (September 13, 2013), https://science.sciencemag.org/content/341/6151/1235367.

166 "The future I see": Personal communication, February 15, 2019.

169 "Post–World War II": Personal communication, February 8, 2019.

170 "I'm probably wrong": Personal communication, February 8, 2019.

172 "stubbornly optimistic": Personal communication, October 15, 2019.

180 "a radically utopian way": Holly Jean Buck, "The Need for Carbon Renewal," *Jacobin*, July 24, 2018, https://jacobinmag.com/2018/07/carbon-removal-geoengineering-global-warming.

182 "potentially delusional assumptions": Pablo Suarez, "Geoengineering: A Humanitarian Concern," *Earth's Future* 5, no. 2 (December 23, 2016): 183–95, https://agupubs.onlinelibrary.wiley.com/doi/10.1002/2016EF000464.

182 "If a hundred countries": "Climate Futures—The Rise of Geoengineering and Its Potential Impacts for the Humanitarian Sector,"

IFRC, February 20, 2017, https://www.youtube.com/watch?v=2oVnasx6hAo.

184 "The best-case scenario": Personal communication, February 8, 2019.

189 "Instead of trying to force climate change": Personal communication, February 20, 2019.

189 "Climate change isn't something": Personal communication, February 15, 2019.

190 "We should really start": Personal communication, February 19, 2019.

**Epilogue**

196 "one of the greatest national crises": "Hurricane Dorian Wipes Out Parts of Bahamas," CNN Newsroom, transcript, September 4, 2019, http://us.cnn.com/TRANSCRIPTS/1909/04/cnr.18.html.

197 "not because they are easy": John F. Kennedy's Inspirational Speech: "We Choose to Go to the Moon," September 12, 1962, video posted July 26, 2010, YouTube, stissi101, https://www.youtube.com/watch?v=Ateh7hnEnik.

199 "The death toll": Bernard Ferguson, "Hurricane Dorian Was a Climate Injustice," *The New Yorker*, September 12, 2019, https://www.newyorker.com/news/news-desk/hurricane-dorian-was-a-climate-injustice.

202 "The reality is": Personal communication, May 19, 2016.

205 A Grief Exercise: Contributed by Caroline Contillo.

217 An Imagination Exercise: Contributed by Lawrence Barriner II.